Aamir Al Mosawi

L-arginine research progress

D1798991

Aamir Al Mosawi

L-arginine research progress

L-arginine research

LAP LAMBERT Academic Publishing

Impressum/Imprint (nur für Deutschland/only for Germany)
Bibliografische Information der Deutschen Nationalbibliothek: Die Deutsche
Nat onalbibliothek verzeichnet diese Publikation in der Deutschen Nationalbibliografie;
detaillierte bibliografische Daten sind im Internet über http://dnb.d-nb.de abrufbar.
Alle in diesem Buch genannten Marken und Produktnamen unterliegen warenzeichen-,
marken- oder patentrechtlichem Schutz bzw. sind Warenzeichen oder eingetragene
Warenzeichen der jeweiligen Inhaber. Die Wiedergabe von Marken, Produktnamen,
Gebrauchsnamen, Handelsnamen, Warenbezeichnungen u.s.w. in diesem Werk berechtigt
auch ohne besondere Kennzeichnung nicht zu der Annahme, dass solche Namen im Sinne
der Warenzeichen- und Markenschutzgesetzgebung als frei zu betrachten wären und
daher von jedermann benutzt werden dürften.

Coverbild: www.ingimage.com

Verlag: LAP LAMBERT Academic Publishing GmbH & Co. KG
Heinrich-Böcking-Str. 6-8, 66121 Saarbrücken, Deutschland
Telefon +49 681 3720-310, Telefax +49 681 3720-3109
Email: info@lap-publishing.com

Herstellung in Deutschland (siehe letzte Seite)
ISBN: 978-3-8484-9807-9

Imprint (only for USA, GB)
Bibl ographic information published by the Deutsche Nationalbibliothek: The Deutsche
Nationalbibliothek lists this publication in the Deutsche Nationalbibliografie; detailed
bibl ographic data are available in the Internet at http://dnb.d-nb.de.
Any brand names and product names mentioned in this book are subject to trademark,
brand or patent protection and are trademarks or registered trademarks of their respective
holders. The use of brand names, product names, common names, trade names, product
descriptions etc. even without a particular marking in this works is in no way to be
construed to mean that such names may be regarded as unrestricted in respect of
trademark and brand protection legislation and could thus be used by anyone.

Cover image: www.ingimage.com

Pub isher: LAP LAMBERT Academic Publishing GmbH & Co. KG
Heinrich-Böcking-Str. 6-8, 66121 Saarbrücken, Germany
Pho ne +49 681 3720-310, Fax +49 681 3720-3109
Email: info@lap-publishing.com

Printed in the U.S.A.
Printed in the U.K. by (see last page)
ISBN: 978-3-8484-9807-9

L-arginine research progress

Professor, Aamir Jalal Al Mosawi M.D, Ph.D.
1-Advisor Doctor
Training and Development Center
Iraqi Ministry of Health
2- Head Iraq Headquarter of Copernicus Scientists International Panel
PO Box 70025, Baghdad, Iraq.
e-mail:almosawiAJ@yahoo.com

Content	3

Preface

L-arginine is an amino acid, involved in diverse physiological and pathological processes. In mammals, arginine is classified as a semi-essential or conditionally essential amino acid, depending on the developmental stage and health status of the individual. Arginine plays an important role in cell division, the healing of wounds, removing ammonia from the body, immune function, and the release of hormones

Increasing evidence implicates L-arginine in the pathogenesis of diverse age-related diseases, including atherosclerosis, ischemic heart diseases, renal diseases, diabetes, inflammation, erectile dysfunction and Alzheimer's disease. Researches suggested that L-Arginine can be therapeutically useful in a variety of conditions. Researches suggested the possibility of using L-arginine to control the cardiovascular risk factors in the primary prevention of atherosclerosis and its clinical outcomes before development of irreversible vascular damage. In addition researches suggested that supplementation of L-arginine along with regular therapy may be beneficial to the patients with ischemic myocardial syndromes.

L-arginine allows acceleration or attenuation of some metabolic steps contributing to a valuable course to neurogenerative therapies for Alzheimer's disease. L-arginine supplement prevents vascular oxidative stress and inflammation in a murine model of T1D diabetes. Some studies provided the rationale for novel therapies, including supplementation of dietary L-arginine or its precursor L-citrulline, inhibition of non-NO producing pathways of L-arginine utilization, or both in end-stage renal disease. The benefits and functions attributed to oral supplementation of L-arginine include:

1-Precursor for the synthesis of nitric oxide (NO)
2-Reduces healing time of injuries (particularly bone)
3-Quickens repair time of damaged tissue
4-Helps decrease blood pressure

Understanding the potential roles of L-arginine may lead to the rational development of therapeutic agents for various diseases. L-Arginine is responsible for the production of nitric oxide in the body which is a proven key molecule in relaxing blood vessels and improving circulation.

1- L-Arginine

L-Arginine is a semi-essential α-amino acid involved in numerous areas of human physiology, including production of nitric oxide (NO) – a key messenger molecule involved in vascular regulation, immune activity, and endocrine function. Arginine is also involved in protein production, wound healing, erectile function, and fertility. Arginine is not considered essential because humans can synthesize it de novo from glutamine, glutamate, and proline. In mammals, arginine is classified as a semi-essential or conditionally essential amino acid, depending on the developmental stage and health status of the individual. Preterm infants are unable synthesize or create arginine internally, making the amino acid nutritionally essential for them [1-11].

Dietary intake remains the primary determinant of plasma arginine levels, since the rate of arginine biosynthesis does not compensate for depletion or inadequate supply. Individuals who have poor nutrition or certain physical conditions may be advised to increase their intake of foods containing arginine. Arginine is found in a wide variety of foods listed in Table 1:1[4].

Animal sources
Dairy products (e.g. cheese, milk, yogurt, whey protein drinks), beef, pork (e.g. bacon, ham), gelatin , poultry (e.g. chicken and turkey light meat), pheasant, quail, seafood (e.g. halibut, lobster, salmon, shrimp, snails, tuna).
Plant sources
Wheat germ and flour, buckwheat, oat meal, peanuts, nuts (coconut, pecans, cashews, walnuts, almonds, Brazil nuts, hazelnuts, seeds (pumpkin, sesame, sunflower), chick peas, cooked soybeans.

Table (1:1) Dietary sources of Arginine

Arginine is the most abundant nitrogen carrier in humans, containing four nitrogen atoms per molecule. Arginine is not a major inter-organ nitrogen shuttle; instead, it plays an important role in nitrogen metabolism and ammonia detoxification as an intermediate in the urea cycle [8, 9, 10]

Arginine is synthesized in mammals from glutamine via pyrroline carboxyl ate (P5C) synthase and proline oxidase in a multi-step metabolic conversion. In adults, most endogenous arginine is produced from citrulline, a by-product of glutamine metabolism in the gut and liver. Citrulline is released into the circulation and taken up primarily by the kidney for conversion into arginine [11, 12]

The amino acid side chain of arginine consists of a 3-carbon aliphatic straight chain, the distal end of which is capped by a complex guanidinium group. The L-form is one of the 20 most common natural amino acids. At the level of molecular genetics, in the structure of the messenger ribonucleic acid mRNA, CGU, CGC, CGA, CGG, AGA, and AGG, are the triplets of nucleotide bases or cordons that codify for arginine during protein synthesis. [1, 2].

Arginine is synthesized from citrulline by the sequential action of the cytosolic enzymes argininosuccinate synthetase (ASS) and argininosuccinate lyase (ASL). This is energetically costly, as the synthesis of each molecule of argininosuccinate requires hydrolysis of adenosine triphosphate (ATP) to adenosine monophosphate (AMP); *i.e.*, two ATP equivalents.

Citrulline can be derived from multiple sources:

- from arginine via nitric oxide synthase (NOS)
- from ornithine via catabolism of proline or glutamine/glutamate
- from asymmetric dimethylarginine (ADMA) via DDAH

The pathways linking arginine, glutamine, and proline are bidirectional. Thus, the net utilization or production of these amino acids is highly dependent on cell type and developmental stage.

On a whole-body basis, synthesis of arginine occurs principally via the intestinal–renal axis, wherein epithelial cells of the small intestine, which produce citrulline primarily from glutamine and glutamate, collaborate with the proximal tubule cells of the kidney, which extract citrulline from the circulation and convert it to arginine, which is returned to the circulation.

Consequently, impairment of small bowel or renal function can reduce endogenous arginine synthesis, thereby increasing the dietary requirement.

Synthesis of arginine from citrulline also occurs at a low level in many other cells, and cellular capacity for arginine synthesis can be markedly increased under circumstances that also induce NOS. Thus, citrulline, a co-product of the NOS-catalyzed reaction, can be recycled to arginine in a pathway known as the citrulline-NO or arginine-citrulline pathway. This is shown the fact that in many cell types, citrulline can substitute for arginine to some degree in supporting NO synthesis. However, recycling is not quantitative because citrulline accumulates along with nitrate and nitrite, the stable end-products of NO, in NO-producing cells [5].

The benefits and functions attributed to oral supplementation of L-arginine include [6- 15]:

1-Precursor for the synthesis of nitric oxide (NO).
2-Reduces healing time of injuries (particularly bone).
3-Quickens repair time of damaged tissue.
4-Helps decrease blood pressure.

Arginine is the immediate precursor of NO, urea, ornithine and agmatine; is necessary for the synthesis of creatine; and can also be used for the synthesis of polyamines (mainly through ornithine and to a lesser degree through agmatine), citrulline, and glutamate. As a precursor of nitric oxide, arginine may have a role in the treatment of some conditions where vasodilatation is required[2] The presence of asymmetric dimethylarginine (ADMA), a close relative, inhibits the nitric oxide reaction; therefore, ADMA is considered a marker for vascular disease, just as L-arginine is considered a sign of a healthy endothelium.

Supplemental arginine is readily absorbed.6 About 50-percent of ingested arginine is rapidly converted in the body to ornithine, primarily by the enzyme arginase. Because of this fast turnover, sustained-release preparations are being investigated as a way to maintain a steadier blood level over time. Ornithine, in turn, can be metabolized to glutamate and proline, or through the enzyme ornithine decarboxylase into the polyamine pathway for degradation into compounds such as putrescine and other polyamines [16, 17].

Arginine is a precursor for the synthesis of nitric oxide, proteins, urea, creatine, vasopressin, and agmatine [18]. Arginine that is not metabolized by arginase to ornithine is processed by one of four other enzymes: nitric oxide synthase (to become nitric oxide); arginine: glycine amidinotransferase (to become creatine); arginine decarboxylase (to become agmatine); or arginyl-tRNA synthetase (to become arginyl-tRNA, a precursor to protein synthesis). Arginine is also an allosteric activator of N-acetylglutamate synthase, which synthesizes N-acetylglutamate
from glutamate and acetyl-CoA [19].

References

1-Tapiero, H.; et al. (November 2002). "L-Arginine". Biomedicine and Pharmacotherapy 56 (9): 439–445.

2-Wu, G.; et al. (August 2004). "Arginine deficiency in preterm infants: biochemical mechanisms and nutritional implications". Journal of Nutritional Biochemistry 15 (8): 332–451.

3-"L-Arginine Supplements Nitric Oxide Scientific Studies Food Sources". http://www.keysupplements.com/articles/L-Arginine-Supplements-Nitric-Oxide-Scientific-Studies.htm. Accessed 2011-09-2.

4- Morris Jr SM. "Enzymes of arginine metabolism.". The Journal of nutrition 2004; 134 (10 Suppl): 2743S–2747S.

5-Stechmiller JK, et al. "Arginine supplementation and wound healing". Nutrition in Clinical Practice 2005; 20 (13): 52–61.

6-Witte MB, Barbul A. "Arginine physiology and its implication for wound healing". Wound Repair and Regeneration 2003; 11 (6): 419–423.

7-Andrew PJ, Myer B. "Enzymatic function of nitric oxide synthases". Cardiovascular Research 1999; 43 (3): 521–531. 8- Castillo L, Chapman TE, Sanchez M, et al. Plasma arginine and citrulline kinetics in adults given adequate and arginine-free diets. Proc Natl Acad Sci U S A 1993; 90:7749-7753.

9- Castillo L, Ajami A, Branch S, et al. Plasma arginine kinetics in adult man: response to an arginine-free diet. Metabolism 1994; 43:114-122.

10- Abcouwer SF, Souba WW. Glutamine and arginine. In: Shils ME, Olson JA, Shike M, Ross AC, eds. Modern Nutrition in Health and Disease, 9 ed. Baltimore, MD: Williams & Wilkins; 1999:559-569.

11- Wu G, Davis PK, Flynn NE, et al. Endogenous synthesis of arginine plays an important role in maintaining arginine homeostasis in postweaning growing pigs. JNutr 1997; 127:2342-2349.

12-Dhanakoti SN, Brosnan JT, Herzberg GR, Brosnan ME. Renal arginine synthesis: studies in vitro and in vivo. Am J Physiol 1990; 259:E437-E442.

13-Gokce N. "L-Arginine and hypertension". Journal of Nutrition 2004; 134 (10 Suppl): 2807S–2811S.

14-Rajapakse NW. et al. "Exogenous L-arginine ameliorates angiotensin II-induced hypertension and renal damage in rats". Hypertension 2008; 52 (6): 1084–1090.

15-Schulman SP, Becker LC, Kass DA, Champion HC, Terrin ML, Forman S, Ernst KV, Kelemen MD et al. "L-arginine therapy in acute myocardial infarction: the Vascular Interaction With Age in Myocardial Infarction (VINTAGE MI) randomized clinical trial.". JAMA: the journal of the American Medical Association 2006; 295 (1): 58–64.

16-Preiser JC, Berre PJ, Van Gossum A, et al. Metabolic effects of arginine addition to the enteral feeding of critically ill patients. JPEN J Parenter Enteral Nutr 2001; 25:182-187.

17- Castillo L, Sanchez M, Vogt J, et al. Plasma arginine, citrulline, and ornithine kinetics in adults, with observations on nitric oxide synthesis. Am J Physiol 1995; 268:E360-E367.16-Wu G, Morris SM Jr. Arginine metabolism: nitric oxide and beyond. Biochem J 1998; 336:1-17.

18-Wu G, Morris SM Jr. Arginine metabolism: nitric oxide and beyond. Biochem J 1998; 336:1-17.
19-Meijer AJ, Lamers WH, Chamuleau RA. Nitrogen metabolism and ornithine cycle function. Physiol Rev 1990; 70:701-748.

2-Mechanisms of Action of L-arginine: An overview

Arginine is the biological precursor of nitric oxide (NO); an endogenous gaseous messenger molecule involved in a variety of endothelium-dependent physiological effects in the cardiovascular system. Much of arginine 's influence on the cardiovascular system is due to endothelial NO synthesis, which results in vascular smooth muscle relaxation and subsequent vasodilatation, as well as inhibition of monocyte adhesiveness, platelet aggregation, and smooth muscle proliferation. A great deal of research has explored the biological roles and properties of nitric oxide, which is also of critical importance in maintenance of normal blood pressure, myocardial function, inflammatory response, apoptosis, and protection against oxidative damage[1-8].

Arginine is a potent immunomodulator. Supplemental arginine appears to up-regulate immune function and reduce the incidence of postoperative infection. Significant decreases in cell adhesion molecules and pro-inflammatory cytokine levels have also been observed. Arginine supplementation (30 g/day for three days) has been shown to significantly enhance natural killer (NK) cell activity; lymphokine activated killer cell cytotoxicity, and lymphocyte mitogenic reactivity in patients with locally advanced breast cancer [9, 10].

Arginine has significant effects on endocrine function – particularly adrenal and pituitary secretion – in humans and animals. Arginine administration can stimulate the release of catecholamines, insulin and glucagon, prolactin, and growth hormone (GH).Little is known about the specific mechanism(s) by which arginine exerts these effects[11,15].

References

1- Wu G, Meininger CJ. Arginine nutrition and cardiovascular function. J Nutr 2000; 130:2626-2629.

2- Gross SS, Wolin MS. Nitric oxide: pathophysiological mechanisms. Annu Rev Physiol 1995; 57:737-769.

3- Wink DA, Hanbauer I, Grisham MB, et al. Chemical biology of nitric oxide: regulation and protective and toxic mechanisms. Curr Top Cell Regul 1996; 34:159-187.

4- Umans JG, Levi R. Nitric oxide in the regulation of blood flow and arterial pressure. Annu Rev Physiol 1995; 57:771-790.

5- Hare JM, Colucci WS. Role of nitric oxide in the regulation of myocardial function. Prog Cardiovasc Dis 1995; 38:155-166.

6- Lyons CR. The role of nitric oxide in inflammation. Adv Immunol 1995; 60:323-371.

7- Brune B, Messmer UK, Sandau K. The role of nitric oxide in cell injury. Toxicol Lett 1995; 82-83:233-237.

8- Wink DA, Cook JA, Pacelli R, et al. Nitric oxide (NO) protects against cellular damage by reactive oxygen species. Toxicol Lett 1995; 82-83:221-226.

9- Brittenden J, Heys SD, Ross J, et al. Natural cytotoxicity in breast cancer patients receiving neoadjuvant chemotherapy: effects of L-arginine supplementation. Eur J Surg Oncol 1994; 20:467-472.

10- Brittenden J, Park KGM, Heys SD, et al. L-arginine stimulates host defenses in patients

11- Imms FJ, London DR, Neame RL. The secretion of catecholamines from the adrenal gland following arginine infusion in the rat. J Physiol 1969; 200:55P-56P.

12- Palmer JP, Walter RM, Ensinck JW. Arginine-stimulated acute phase of insulin and glucagon secretion. I. In normal man. Diabetes 1975; 24:735-740.

13- Rakoff JS, Siler TM, Sinha YN, Yen SS. Prolactin and growth hormone release in response to sequential stimulation by arginine and synthetic TRF. J Clin Endocrinol Metab 1973; 37:641-644.

14- Knopf RF, Conn JW, Fajans SS, et al. Plasma growth hormone response to intravenous administration of amino acids. J Clin Endocrinol Metab 1965; 25:1140-1144.

15- Merimee TJ, Lillicrap DA, Rabinowitz D. Effect of arginine on serum-levels of human growth-hormone. Lancet 1965; 2:668-670.

3-Preventive role of L-arginine in atherosclerosis

Atherosclerosis is the single most important cause of cardiovascular disease (CVD) and a major health problem worldwide [1]. Clinical manifestations of atherosclerosis include myocardial infarction, heart failure, stroke, and peripheral artery disease, result in irreversible organ damage [2]. Early atherosclerosis lesions or fatty streaks become increasingly prevalent among children and young adults in industrialized countries [3, 4]. Fatty streaks lesion formation may begin even before birth as intimal thickening can be observed in fetal coronary arteries [5]. Although these lesions may be vanishing, some of these lesions progress to advanced stages of atherosclerosis. As atherosclerosis has a long asymptomatic phase and the first manifestation of disease may be sudden cardiac death, it is imperative to find effective strategies to prevent it [4, 6].

Current guidelines for the prevention of atherosclerotic diseases focus on treatment of established cardiovascular risk factors to attenuate the subsequent endothelial cell dysfunction and damage [6]. Endothelial dysfunction (ED) is an early event in atherosclerosis and has a pivotal role in the atherogenesis process [7].

L-Arginine (2-amino-5-guanidinovaleric acid) is the precursor of nitric oxide, an endogenous messenger molecule involved in a variety of endothelium-mediated physiological effects in the vascular system. Acute and chronic administration of L-Arginine has been shown to improve endothelial function in animal models of hypercholesterolemia and atherosclerosis. L-Arginine also improves endothelium-dependent vasodilatation in humans with hypercholesterolemia and atherosclerosis. The responsiveness to L-Arginine depends on the specific cardiovascular disease studied, the vessel segment, and morphology of the artery. The pharmacokinetics of L-Arginine has recently been investigated. Side effects are rare and mostly mild and dose dependent. The mechanism of action of L-Arginine may involve nitric oxide synthase substrate provision, especially in patients with elevated levels of the endogenous NO synthase inhibitor asymmetric dimethylarginine. Endocrine effects and unspecific reactions may contribute to L-Arginine- induced vasodilatation after higher doses. Several long-term studies have been performed that show that chronic oral administration of L-Arginine or intermittent infusion therapy with L-Arginine can improve clinical symptoms of cardiovascular disease in man.

ED is an independent predictor of cardiovascular events [6, 7]. The vascular endothelium is a crucial regulator of vascular function and homeostasis. Nitric oxide (NO) is an important paracrine substance released by the endothelium to regulate vasomotor tone. Risk factors for atherosclerosis, as well as atherosclerosis per se, are associated with endothelial dysfunction and decreased bioavailability of NO.

ED is a reversible disorder and search for the proper time before development of irreversible vascular injury is extremely important [4]. ED is characterized by reduced bioavailability of nitric oxide (NO) [8-11]. NO is a potent anti-atherosclerotic molecule and every intervention that enhances NO bioavailability might be a promising strategy for the prevention and treatment of atherosclerosis [10-13]. One straightforward approach to increase NO bioavailability is providing additional substrate for nitric oxide synthase (NOS). L-Arginine is the substrate of endothelial NOS (eNOS) and the main precursor of NO in the vascular endothelium. Data from numerous studies imply that L-arginine supplementation restores endothelial function in several disease states associated with ED such as hypercholesterolemia [14-18]. Despite the positive results from several studies, there are some studies that have shown that L-arginine administration did not improve endothelium- dependent dilation or the inflammatory state of patients [18]. L-arginine could even be harmful to vascular health; for example L-arginine supplementation during the post-myocardial-infarction period was associated with higher post-infarction mortality than placebo [19]. So, it is still unclear whether administration of L-arginine has any beneficial effect on clinical outcome [18].

Several studies investigated L-arginine beneficial effects on endothelial function, particularly in hypercholesterolemia animals, have supplemented L-arginine during induction of hypercholesterolemia [17, 26-28]. These experiments showed that ingestion of L-arginine, as a boosting supplement for NO production can reverse the state of oxidative stress and progression of atherosclerosis [20].

Dietary supplementation with L-arginine significantly improves EDD in hypercholesterolemic young adults, and this may impact favorably on the atherogenic process.

In hypercholesterolemic rabbits, oral L-arginine attenuates endothelial dysfunction and atheroma formation. Using high resolution external ultrasound, arterial physiology was studied in 27 hypercholesterolemic subjects aged 29+/-5 (19-40) years, with known endothelial dysfunction and LDL-cholesterol levels of 238+/-43 mg/dl. Each subject was studied before and after 4 wk of L-arginine (7 grams x 3/day) or placebo powder, with 4 wk washout, in a randomized double-blind crossover study. Brachial artery diameter was measured at rest, during increased flow (causing endothelium-dependent dilation, EDD) and after sublingual glyceryl trinitrate (causing endothelium-independent dilation). After oral L-arginine, plasma L-arginine levels rose from 115+/-103 to 231+/-125 micromol/liter (P<0.001), and EDD improved from 1.7+/-1.3 to 5.6+/-3.0% (P<0.001). In contrast there was no significant change in response to glyceryl trinitrate. After placebo there were no changes in endothelium-dependent or independent vascular responses. Lipid levels were unchanged after L-arginine and placebo [21].

References

1. Murray CJ, Lopez AD: Global mortality, disability, and the contribution of risk factors: Global Burden of Disease Study. Lancet 1997, 349(9063):1436-1442.
2. Werner N, Nickenig G: Influence of cardiovascular risk factors on endothelial progenitor cells: limitations for therapy? Arterioscler Thromb Vasc Biol 2006, 26(2):257-266.
3. Berenson GS, Srinivasan SR: Prevention of atherosclerosis in childhood. Lancet 1999, 354(9186):1223-1224.
4. Napoli C, Lerman LO, de Nigris F, Gossl M, Balestrieri ML, Lerman A: Rethinking primary prevention of atherosclerosis-related diseases. Circulation 2006, 114(23):2517-2527.
5. Napoli C, D'Armiento FP, Mancini FP, Postiglione A, Witztum JL, Palumbo G, Palinski W: Fatty streak formation occurs in human fetal aortas and is greatly enhanced by maternal hypercholesterolemia. Intimal accumulation of low density lipoprotein and its oxidation precede monocyte recruitment into early atherosclerotic lesions. J Clin Invest 1997, 100(11):2680-2690.
6. Mensah GA, Ryan US, Hooper WC, Engelgau MM, Callow AD, Kapuku GK, Mantovani A: Vascular endothelium summary statement II: Cardiovascular disease prevention and control. Vascul Pharmacol 2007, 46(5):318-320.
7. Bonetti PO, Lerman LO, Lerman A: Endothelial dysfunction: a marker of atherosclerotic risk. Arterioscler Thromb Vasc Biol 2003, 23(2):168-175.
8. Ignarro LJ, Napoli C: Novel features of nitric oxide, endothelial nitric oxide synthase, and atherosclerosis. Curr Diab Rep 2005, 5(1):17-23.
9. Maxwell AJ: Mechanisms of dysfunction of the nitric oxide pathway in vascular diseases. Nitric Oxide 2002, 6(2):101-124.
10. Napoli C, de Nigris F, Williams-Ignarro S, Pignalosa O, Sica V, Ignarro LJ: Nitric oxide and atherosclerosis: an update. Nitric Oxide 2006, 15(4):265-279.
11. Rassaf T, Kleinbongard P, Kelm M: The L-arginine nitric oxide pathway: avenue for a multiple-level approach to assess vascular function. Biol Chem 2006, 387(10–11):1347-1349.
12. Megson IL, Webb DJ: Nitric oxide donor drugs: current status and future trends. Expert Opin Investig Drugs 2002, 11(5):587-601.
13. Miller MR, Megson IL: Recent developments in nitric oxidedonor drugs. Br J Pharmacol 2007, 151(3):305-321.
14. Boger RH, Bode-Boger SM, Brandes RP, Phivthong-ngam L, Bohme M, Nafe R, Mügge A, Frölich JC: Dietary L-arginine reduces the progression of atherosclerosis in cholesterol-fed rabbits: comparison with lovastatin. Circulation 1997, 96(4):1282-1290.
15. Cooke JP, Tsao PS: Arginine: a new therapy for atherosclerosis? Circulation 1997, 95(2):311-312.
16. Gornik HL, Creager MA: Arginine and endothelial and vascular health. J Nutr 2004, 134(Suppl 10):2880S-2887S.
17. Hayashi T, Juliet PA, Matsui-Hirai H, Miyazaki A, Fukatsu A, Funami J, Iguchi A, Ignarro LJ: l-Citrulline and l-arginine supplementation retards the progression of high-cholesterol-diet-induced atherosclerosis in rabbits. Proc Natl Acad Sci USA 2005, 102(38):13681-13686.

18. Tousoulis D, Boger RH, Antoniades C, Siasos G, Stefanadi E, Stefanadis C: Mechanisms of disease: L-arginine in coronary atherosclerosis – a clinical perspective. Nat Clin Pract Cardiovasc Med 2007, 4(5):274-283.

19. Schulman SP, Becker LC, Kass DA, Champion HC, Terrin ML, Forman S, Ernst KV, Kelemen MD, Townsend SN, Capriotti A, Hare JM, Gerstenblith G: L-arginine therapy in acute myocardial infarction: the Vascular Interaction With Age in Myocardial Infarction (VINTAGE MI) randomized clinical trial. JAMA 2006 295(1):58-64.

20-Javanmard SH, Nematbakhsh M, Mahmoodi F, Mohajeri MR: l- Arginine supplementation enhances eNOS expression in experimental model of hypercholesterolemic rabbits aorta. Pathophysiology 2009.

21- Clarkson P, Adams MR, Powe AJ, et al. Oral L-Arginine Improves Endothelium-Dependent Dilation in Hypercholesterolemic Young Adults. J Am Coll Cardiol. 1997 Mar 1; 29(3):491-7.

4-Role of Arginine in cardiovascular conditions

L-Arginine (2-amino-5-guanidinovaleric acid) is the precursor of nitric oxide, an endogenous messenger molecule involved in a variety of endothelium-mediated physiological effects in the vascular system. Acute and chronic administration of L-Arginine has been shown to improve endothelial function in animal models of hypercholesterolemia and atherosclerosis. L-Arginine also improves endothelium-dependent vasodilatation in humans with hypercholesterolemia and atherosclerosis. The responsiveness to L-Arginine depends on the specific cardiovascular disease studied, the vessel segment, and morphology of the artery. Several long-term studies have been performed that show that chronic oral administration of L-Arginine or intermittent infusion therapy with L-Arginine can improve clinical symptoms of cardiovascular disease in man. Side effects are rare and mostly mild and dose dependent.

Arginine's effects on cardiovascular function are due to arginine-induced endothelial NO production. Endothelial nitric oxide synthase (eNOS) catalyzes this reaction, which produces NO and ornithine. Nitric oxide diffuses into the underlying smooth muscle and stimulates guanylyl cyclase, producing guanosine-3', 5'-cyclic monophosphate (cGMP), which in turn causes muscle relaxation and vasodilatation.

Arginine supplementation has been shown to increase flow-mediated brachial artery dilation in normal individuals as well as those with hyperlipidemia and hypertension. Nitric oxide is also responsible for creating an environment in the endothelium that is anti-atherogenic. Adequate NO production inhibits processes at the core of the atherosclerotic lesion, including platelet aggregation, monocyte adhesion and migration, smooth muscle proliferation, and vasoconstriction [1, 2].

Asymmetrical dimethylarginine (ADMA) competes with arginine for binding with eNOS, subsequently down-regulating activity of this vital enzyme. Increased plasma ADMA has been shown to be an independent risk factor for cardiovascular disease because of its inhibitory activity on eNOS. Oral arginine supplementation overrides the inhibitory effect of ADMA on eNOS, and improves vascular function in those with high ADMA levels [3-5].

Ischemic heart disease can result from a variety of pathophysiological pathways. Regardless the cause of ischemia, the consequence is always decreased oxygen availability to the myocardium.

Insufficient oxygen supply during cardiac ischemia leads to the following important consequences [6-13]:

1- Failure of oxidative phosphorylation to meet the energy demand leading to stimulation of the glycolytic pathway with resultant accumulation of lactic acid resulting in acidosis.

2-Anaerobic production of ATP is insufficient to meet the energy demand of the tissues. Fall in the level ATP initiates a series of events, including oxidative stress conditions, which are deleterious to the endothelial cells.

3-Oxidative stress, associated with increased formation of reactive oxygen species (ROS), modifies phospholipids and proteins leading to lipid per-oxidation and oxidation of thiol groups. These changes are considered to alter the membrane permeability and configuration.

4-Oxidative stress may result in:
A- Depression in the sarcolemmal (SL) Ca^{2+} pump.
B- Depression in ATPase and Na^+-K^+ ATPase activities.
C- The changes in (a and c) lead to decreased Ca^{2+} efflux and increased Ca^{2+} influx and thus inhibit Ca^{2+} sequestration from the cytoplasm in the cardiomyocytes.

5-The depression in Ca^{2+} regulatory mechanism by ROS ultimately results in intracellular Ca^{2+} overload and cell death and loss of heart contractile function and severe myocardial cell damage.

6- An increase in Ca^{2+} during ischemia induces conversion of xanthine dehydrogenase to xanthine oxidase, by selective proteolysis that generates more of super oxide radicals.

7-The super oxide radicals, though less toxic by itself, triggers the formation of other reactive oxygen species. These include OH, H_2O_2 and $HOCl^-$. The hydroxyl radicals, in particular, interact with lipids, proteins and nucleic acids resulting in the loss of membrane integrity, structural and functional changes in enzymes, proteins and genetic material.

The oxidative stress induced changes in sarcoplasmic reticulum (SR) Ca2+ and SL Na+-K+ pumps are not limited only to cardiomyocytes but have also been observed in the coronary artery smooth muscle cells [6].

The body defense mechanisms to deactivate/destroy ROS before they can cause damage include:

1-The body enzymatic free radical scavengers which convert ROS to less toxic or non toxic products:

A-Superoxide dismutase (SOD) SOD is considered to be the first line of defense against oxidative insult, as this enzyme efficiently converts oxygen derived free radical to hydrogen peroxide. Hydrogen peroxide is more toxic than oxygen derived free radicals and is capable of producing most toxic OHo radical. It has, therefore, to be removed very efficiently.

B-Catalase is the highly reactive enzyme in most of the tissues that converts toxic H2O2 to water.

C- Glutathione peroxidase (GPx) removes H2O2 is glutathione peroxidase. This enzyme, almost equally efficient as catalase, removes H2O2 at the expense of oxidation of glutathione. Oxidized glutathione is reduced back by glutathione peroxidase.

D-Glutathione reductase (GR).

Catalase and glutathione peroxidase cooperate in removing H2O2 in many tissues. For example, mammalian red blood cells (RBCs) have both these enzymes. The normal low rate of production of H2O2 by hemoglobin and SOD seems mainly dealt with glutathione peroxidase. However, catalase also contributes to removal of H2O2 if it's concentration is raised [14].

The oxidative damage will occur when the defense mechanisms are deficient, made less active or production of ROS exceeds the capabilities of defense mechanism or a combination of all these.

2-Antioxidant compounds present in the body such as thiols, or taken from outside such as α-tocopherol, β-carotenoids, flavonoids and L-arginine.

The administration of vitamin C[15,16] and vitamin E[17,18] to the patients with ischemic myocardial syndromes lower lipid per-oxidation, increase the activities of free radical scavenging enzymes, which are decreased in these patients, and lower the activity of pro-oxidant enzyme which are elevated in the patients. Oxidized low density lipo-proteins are involved in the progression of atherosclerotic lesions [19] which can be prevented by L-arginine administration. Moreover there are evidences that this amino acid preserves arterial vasodilatation even in the presence of oxidative stress. [20-22].A fine balance between ROS and various anti-oxidant mechanisms is crucial for avoiding myocardial injury.

During ischemia, the myocardial tissue is exposed to an exacerbated injury caused by un-scavenged superoxide radicals, other free radicals, increased WBC, vasoconstriction and platelet adhesion [23-25] Higher oxidant stress and diminished antioxidant status along with higher lmalondialdehyde levels constitutes the key factors in the progression of ischemic injury.

L-arginine as a precursor of nitric oxide (NO) synthesis has the following advantages:

1- NO can reduce oxidative stress by inhibiting XO, scavenge superoxide radicals and terminate free radical chain reaction within the lipid membranes, thereby reducing inflammatory mediators [26, 27]

2-Decreased ROS formation attenuates inhibition of SOD which is increased by L-arginine supplementation. The level of ascorbic acid and total thiols also increased after L-arginine supplementation [26] .The improved levels of ascorbic acid and total thiols cope with lipid and protein per-oxidation thus decreasing the damage of membrane and cellular proteins. By efficiently reducing or eliminating these toxic metabolites L-arginine via NO retards the damaging effects of ischemic injury [27].

3- L-arginine controls the modification of proteins due to excess carbonylation. L-arginine was found to significantly decrease the plasma carbonyl contents in the patients. NO generated by L-arginine also inhibits vasoconstriction, platelet aggregation, expression of adhesion molecules [28], and chemotactic proteins [29].

4-L-arginine administration to the patients has serum cholesterol lowering effect [30, 31].The observed lowering of serum cholesterol upon L-arginine administration has been ascribed to be mediated by NO derived from this amino acid [32].The adhesiveness of mononuclear cells is increased in hyper-cholesterolemic conditions. The increase in adhesiveness and cholesterol levels lead to atherosclerosis and myocardial infarction.

L-arginine supplementation, along with standard therapy, can maintain oxidant-antioxidant homeostasis and control the levels of lipid, cholesterol, ascorbic acid and protein oxidation in patients of myocardial ischemic syndromes and can be taken care of by L-arginine.

It has been shown that L-arginine administration (three grams per day for 15 days) resulted in increased activity of free radical scavenging enzyme superoxide dismutase (SOD) and increase in the levels of total thiols (T-SH) and ascorbic acid with concomitant decrease in lipid per-oxidation, carbonyl content, serum cholesterol and the activity of proxidant enzyme, xanthine oxidase (XO). Therefore, supplementation of L-arginine along with regular therapy may be beneficial to the patients of ischemic heart diseases [33].

Coronary endothelial dysfunction is characterized by an imbalance between endothelium-derived vasodilating and vasoconstricting factors and coronary vasoconstriction in response to the endothelium-dependent vasodilator acetylcholine. A double-blind, randomized study proposed a role for L-arginine as a therapeutic option for patients with coronary endothelial dysfunction and non-obstructive coronary artery disease. Twenty-six patients without significant coronary artery disease on coronary angiography and intravascular ultrasound were blindly randomized to either oral L-arginine or placebo, 3 g TID. After 6 months, the coronary blood flow in response to acetylcholine in the subjects who were taking L-arginine increased compared with the placebo group (149 +/- 20% versus 6 +/- 9%, P < 0.05). This was associated with a decrease in plasma endothelin concentrations and an improvement in patients' symptoms scores in the L-arginine treatment group compared with the placebo group. This study showed that on-term oral L-arginine supplementation for 6 months in humans improves coronary small-vessel endothelial function in association with a significant improvement in symptoms and a decrease in plasma endothelin concentrations [34].

Physiologic coronary vasodilatation correlates with endothelial function and that L-arginine, the substrate for nitric oxide synthesis, improves the response to acetylcholine (Ach).Changes in coronary blood flow and epicardial diameter response to Ach, adenosine and cardiac pacing were measured in 32 patients with coronary atherosclerosis or its risk factors and in 7 patients without risk factors and normal coronary angiograms. Intra-coronary L-arginine did not alter baseline coronary vascular tone, but the epicardial and microvascular responses to Ach were enhanced (both p < 0.001). The improvement after L-arginine was greater in epicardial segments that initially constricted with Ach; similarly, L-arginine abolished microvascular constriction produced by higher doses of Ach. There was a negative correlation between the initial epicardial and vascular resistance responses to Ach and the magnitude of improvement with L-arginine (r = -0.55 and r = -0.50, respectively, p < 0.001). D-Arginine did not affect the responses to Ach, and adenosine responses were unchanged with L-arginine. Cardiac pacing-induced epicardial constriction was abolished by L-arginine, but microvascular dilation remained unaffected. Prevention of epicardial constriction during physiologic stress by L-arginine in patients with endothelial dysfunction may be of therapeutic value in the treatment of myocardial ischemia [35].

Dietary supplementation with vitamin E has been shown to reduce ischemic events in patients with established coronary artery disease and improves endothelial function in cholesterol-fed rabbits. However, a study examined whether such dietary supplementation with vitamin E improves endothelial function in twenty patients with mild hypercholesterolemia and coronary artery disease. The study showed that acute intra-arterial administration of L-arginine (10 mg/min) reversed endothelial dysfunction in forearm vasculature of patients with mild hypercholesterolemia and coronary artery disease but supplementation with vitamin E (400 i.u. daily) for 8 weeks does not reverse L-arginine responsive endothelial dysfunction[36].

Clinical applications

1-Angina Pectoris

It has been suggested that there is a relative deficiency of L-arginine in diseased coronary arteries [37]. Arginine supplementation has been effective in angina treatment in some, but not all, clinical trials. In 36 patients with chronic, stable angina given 6 g arginine daily for two weeks, significant improvement was noted in flow-mediated vasodilatation, exercise time, and quality of life, compared to placebo. No improvement was seen in ischemia markers on ECG or in time-to-onset of angina [38].

In a small, uncontrolled trial, seven of 10 people with intractable angina improved dramatically after taking 9 g arginine daily for three months [39].
A randomized, double-blind, placebo-controlled study in 22 patients with clinical symptoms of stable angina pectoris and healed myocardial infarction has shown that oral supplementation with L-arginine (6 g/day for 3 days) increases exercise capacity (tested on a Marquette case 12 treadmill according to the modified Bruce protocol).The study suggested that the inefficient L-arginine/nitric oxide system contributes to limitation of myocardial perfusion and/or peripheral vasodilatation during maximum exercise in patients with stable angina pectoris[40].

However, in men with stable angina, oral supplementation with arginine (15 g/day) for two weeks was not associated with improvement in endothelium-dependent vasodilatation, oxidative stress, or exercise performance [41].

In patients with coronary artery disease, oral supplementation of arginine (6 g/day for three days) did not affect exercise-induced changes in QT interval duration, QT dispersion, or the magnitude of ST-segment depression; however, it did significantly increase exercise tolerance. The therapeutic effect of arginine in patients with microvascular angina is considered to be the result of improved endothelium-dependent coronary vasodilatation [42, 43].

2- Syndrome X

Syndrome X (angina, normal coronary arteriogram and positive exercise test) remains an enigma with unexplained features. The Syndrome is characterized by generalized flow-related endothelial dysfunction, myocardial thallium 201 defects reflect myocardial or microvascular dysfunction, and endothelial dysfunction and its consequences that may be improved by oral L-arginine [44].

3-Congestive Heart Failure

Infusion of L-Arginine in patients with congestive heart failure results in increased production of nitric oxide, peripheral vasodilatation and increased cardiac output, suggesting a beneficial hemodynamic and possibly therapeutic profile. Twelve patients with congestive heart failure (New York Heart Association class II or III) due to coronary artery disease (left ventricular ejection fraction < 35%) were given 20 g of L-Arginine by intravenous infusion over 1 h at a constant rate. L-Arginine resulted in a significant increase in stroke volume and cardiac output without a change in heart rate. Mean arterial blood pressure decreased (from 102 +/- 11 mm Hg to 89 +/- 9.5 mm Hg, p < 0.002), and systemic vascular resistance decreased significantly. One hour after cessation of L-Arginine infusion, blood pressure, stroke volume, cardiac output and systemic vascular resistance were statistically not different from baseline values. Clearance of NO2/NO3 increased significantly during L-arginine administration [45].

Six weeks of oral arginine supplementation (5.6-12.6 g/d) significantly improved blood flow, arterial compliance, and functional status in patients with congestive heart failure (CHF), compared to placebo, in a randomized, double-blind trial [46]. Another double-blind trial found arginine supplementation (5 g three times daily) improved renal function in people with CHF [47]. After a one-week oral dosing with 6 g arginine daily in 30 males with stable CHF, significant improvements were seen in exercise duration, anaerobic threshold, and VO2[48]. African Americans are at significantly greater risk for development of CHF than Caucasians. However, the improvement in endothelial function seen with arginine dosing may be more pronounced in African Americans compared to Caucasians, as was seen in a study of 52 CHF patients treated with an intra-coronary infusion of arginine [49].

4-Hypertension

Administration of arginine prevented hypertension in salt-sensitive rats, but not in spontaneously hypertensive rats [50]. If arginine was provided early, hypertension and renal failure could be prevented.

In healthy human subjects, intravenous (IV) administration of arginine had vasodilatory and antihypertensive effects [51] In a small, controlled trial, hypertensive patients refractory to enalapril and hydrochlorothiazide responded favorably to the addition of oral arginine (2 g three times daily)[52] Small, preliminary trials have found oral [53] and IV [54] arginine significantly lowers blood pressure in healthy volunteers. IV infusion of arginine (15 mg/kg body weight/min for 35 min) improved pulmonary vascular resistance index and cardiac output in infants with pulmonary hypertension [55].

Endothelium-dependent vasodilatation is attenuated in patients with hypertension. In a prospective randomized double blind trial, 35 patients with essential hypertension received either 6 g L-arginine (18 subjects) or placebo (17 subjects). Administration of L-arginine or placebo did not change significantly heart rate, blood pressure, baseline diameter, blood flow or reactive hyperemia. L-Arginine resulted in a significant improvement of flow-mediated dilatation while placebo did not significantly change this parameter. The effect of L-arginine on flow-mediated dilatation was significantly different from the effect of placebo (P=0.05). L-Arginine did not significantly influence nitrate-induced dilatation. Oral administration of L-arginine acutely improves endothelium-dependent, flow-mediated dilatation of the brachial artery in patients with essential hypertension [56].

In thirteen hypertensive patients with microvascular angina who received oral L-Arginine (4 weeks, 2 g, 3 times daily). L-arginine significantly improved angina class, systolic blood pressure at rest, and quality of life. Maximal forearm blood flow, plasma L-arginine, L-Arginine: asymmetric dimethyl Arginine ratio, and cyclic guanylate monophosphate increased significantly after treatment. This study suggested that in medically treated hypertensive patients with micro-vascular angina, oral L-arginine may represent a useful therapeutic option [57].

5-Intermittent Claudication

Intravenous arginine injections significantly improved symptoms of intermittent claudication in a double-blind trial. Eight grams of arginine, infused twice daily for three weeks, improved pain-free walking distance by 230 ± 63 percent and the absolute walking distance by 155 ± 48 percent (each p <0.05) compared to no improvement with placebo[58].

6-Preeclampsia

Endothelial dysfunction appears to be involved in the pathogenesis of preeclampsia [59]. In an animal model of experimental preeclampsia, IV administration of arginine (0.16 g/kg body weight/day) from gestational day

10 until term reversed hypertension, intrauterine growth retardation, proteinuria, and renal injury [60] Intravenous infusion of arginine (30 g) in preeclamptic women has reportedly increased systemic NO production and reduced blood pressure [61].

Reference

1- Lekakis J, Papathanasieu S, Papamicheel C, et al. Oral l-arginine improves endothelial dysfunction in patients with essential hypertension. J Am Coll Cardiol 2001:260A.

2- Boger GI, Maas R, Schwedhelm E, et al. Improvement of endothelium-dependent vasodilation by simvastatin is potentiated by combination with l-arginine in patients with elevated asymmetric dimethylarginine levels. J Am
Coll Cardiol 2004:525A.

3- Boger RH, Vallance P, Cooke JP. Asymmetric dimethylarginine (ADMA): a key regulator of nitric oxide synthase. Atherosclerosis Suppl 2003; 4:1-3.

4-. Boger RH. Asymmetric dimethylarginine, an endogenous inhibitor of nitric oxide synthase, explains the "L-arginine paradox" and acts as a novel cardiovascular risk factor. J Nutr 2004; 134:2842S-2847S.

5- Boger RH, Ron ES. L-Arginine improves vascular function by overcoming deleterious effects of ADMA, a novel cardiovascular risk factor. Altern Med Rev 2005; 10:14-23.

6-Cerconi C, Cargnoni A, Pasini E, Condorelli E, Curello S, Ferrari R. Evaluation of phospolipid peroxidation as malondialdehyde during myocardial ischemia and reperfusion injury. Am J Physiol 1991; 260:1057-61.

7- Dhalla NS, Panagia V, Singal PK, Makino N, Dixon IM, Eyolfson DA. Alterations in heart membrane calcium transport during the development of ischemia-reperfusion injury. J Mol Cell Cardiol 1988; 20:3-13.

8- Suzuki S, Kaneko M, Chapman DC, Dhalla NS. Alterations in cardiac contractile proteins due to oxygen free radicals. Biochim Biophys Acta 1997; 1074:95-100.

9- Netticadan T, Temsah R, Osada M, Dhalla NS. Status of Ca^{2+}/calmodulin protein kinase phosphorylation of cardiac SR proteins in ischemia-reperfusion. Am J Physiol 1999; 277:384-91.

10 Smart SC, Sagar KB, el Schultz J, Warlteir DC, Jones LR. Injury to the Ca2+ ATPase of the Sarcoplasmic reticulum in anesthesized dogs contributes to myocardial reperfusion injury. Cardiovasc Res 1997; 36:174-84.

11-Elmoselhi AB, Butcher A, Samson SE, Grover AK. Free radicals uncouple the sodium pump in pig coronary artery. Am J Physiol 1994; 266:720-8.

12- Goldhaber JI, Weiss JN. Oxygen free radicals and cardiac reperfusion abnormalities. Hyperten 1992; 20:118-27.

13-Gilham B, Papachristodoulou DK, Thomas JH. Will's biochemical basis of medicine. Butterworth-Heinemann (ed.,) 1997; 3:343-54.

14- Gaetani GF, Kirkman HN, Mangerini R, Ferraris AM. Importance of catalase in the disposal of $H2O2$ within human erythrocytes. Blood 1995; 84:325-30.

15-Bhakuni P, Chandra M, Misra MK. Oxidative stress parameters in erythrocytes of post reperfused patients with myocardial infarction. J Enz Inhibt Med Chem 2005; 20:377-81.

16- Bhakuni P, Chandra M, Misra MK. Effect of ascorbic acid supplementation on certain oxidative stress parameters in the post-reperfusion patients of myocardial infarction. Mol Cell Biochem 2006; 290:153-8.

17- Dwivedi VK, Chandra M, Misra PC, Misra MK. Effect of vitamin E on platelet enzymatic anti-oxidants in the patients of myocardial infarction. Ind J Clin Biochem 2005; 20:21-5.

18- Raghuvanshi R, Chandra M, Misra PC, Misra MK. Effect of vitamin E on the platelet xanthine oxidase and lipid per-oxidation in the patients of myocardial infarction. Ind J Clin   Biochem   2005;   20:26-9.
19- Diaz MN, Frei B, Vita JA, Keaney JF. Anti-oxidants and atherosclerotic heart disease.   N   Engl   J   Med   1997;   337:408-16.
20- Wu G, Meininger CJ. Arginine nutrition and cardiovascular function. J Nutr 2006; 130:2626-9.
21-Bogle RG, Coade SB, Moncada S, Pearson JD, Mann GE. Bradykinin and ATP stimulate L-arginine uptake and nitric oxide release in vascular endothelial cells. Biochem   Biophys   Res   Commun   1991;   180:926-32.
22- Brown AA, Hu FB. Dietary modulation of endothelial function: Implication for cardiovascular   disease.   Am   J   Clin   Nutr   2001;   73:673-86.
23- Buckberg GD. Studies of hypoxemic/reoxygenation injury. Linkage between cardiac function and oxidant damage. J Thorac Cardiovasc Surg 1995; 110:1164-70.
24- Ihnken K, Morita K, Buckberg GD, Sherman MP, Allen BS. Studies of hypoxemic/reoxygenation injury without aortic clamping II. Evidence for reoxygenation damage.   J   Thorac   Cardiovasc   Surg   1995;   110:1171-81.
25-Mizuno A, Baretti R, Buckberg GD, Young H, Vinten-Johansen J, Ma X. Endothelial stunning and myocyte recovery after reperfusion of jeopardized muscle: a role ofL-arginine blood cardioplegia. J Thorac Cardiovasc Surg 1997; 113:379-89.
26-Wan-Teng Lin, Suh-Ching Yang, Shiow-Chwen Tsai, Chi-Chang Huang, Ning-Yuean Lee. L-arginine attenuates xanthine oxidase and myeloperoxidase activities in hearts of rats   during   exhaustive   exercise.   Brit   Jour   Nutr   2006;   95:67-75
27- Engelman DT, Watanabe M, Maulik N, Cordis GA, Engelman RM, Rousou JA, et al.L-arginine reduces endothelial inflammation and myocardial stunning during ischemia/reperfusion.   Ann   Thorac   Surg   1995;   60:1275-81.
28- Theilmeier G, Chan JR, Zalpour S, Anderson B, Wang B. Adhesiveness of MononuclearCells in Hypercholesterolemic Humans is normalized by Dietary L-Arginine.   ArterioscleroThromb   Vasc   Biol   1997;   17:3557-64.
29-Tousoulis D, Boger RH, Antoniades C, Siasos G, Stefanadi E, Stefanadis C. Mechanisms of disease: L-arginine in coronary atherosclerosis—a clinical perspective. Nat   Clin   Pract   Cardiovas   Med   2007;   4:274-83.
30- Rossitch E Jr, Alexander E, Black P, Cooke JP. L-arginine normalizes endothelial function in cerebral vessels from hypercholesterolemic rabbits. J Clin Invest 1991; 87:1295-9.
31-Hurson M, Regan MC, Kirk SJ, Wasserkrug HL. Barbul A.1 Metabolic Effects of Arginine in a Healthy Elderly Population. J Parent Enter Nutr 1995; 19:227-30.
32- Korbut R, Bieron K, Gryglewski RJ. Effect of L-arginine on plasminogen-activator inhibitor in hypertensive patients with hypercholesterolemia. The New Engl J Med 1993; 328:287-8.
33-Tripathi P, Chandra M, Misra MK.Oral administration of L-arginine in patients with angina or following myocardial infarction may be protective by increasing plasma superoxide dismutase and total thiols with reduction in serum cholesterol and xanthine oxidase. Oxidative Medicine and Cellular Longevity 2009; 2:4, 231-237.

34-Lerman A, Burnett JC Jr, Higano ST, McKinley LJ, Holmes DR Jr.Long-Term L-Arginine Supplementation Improves Small-Vessel Coronary Endothelial Function in Humans. Clin Sci (Lond). 1998; 94(2):129-34.

35-Quyyumi AA, Dakak N, Diodati JG, Gilligan DM, Panza JA, Cannon RO 3rd. Effect of L-Arginine on Human Coronary Endothelium-Dependent and Physiologic Vasodilation. Circulation. 1998 Jun 2; 97(21):2123-8.

36-Chowienczyk PJ, Kneale BJ, Brett SE, Paganga G, Jenkins BS, Ritter JM.Lack of Effect of Vitamin E on L-Arginine-Responsive Endothelial Dysfunction in Patients with Mild Hypercholesterolemia and Coronary Artery Disease
 Am J Cardiol. 1997 Aug 1; 80(3):331-3.

37-Tousoulis D, Davies GJ, Tentolouris C, et al. Effects of Changing the Availability of the Substrate for Nitric Oxide Synthaseby L-Arginine Administration on Coronary Vasomotor Tone in Angina Patients with Angiographically Narrowed and in Patients with Normal Coronary Arteries. Atherosclerosis. 1997 Mar 21; 129(2):261-9.

38-Maxwell AJ, Zapien MP, BS, Pearce GL, et al. Randomized trial of a medical food for the dietary management of chronic, stable angina. J Am Coll Cardiol 2002; 39:37-45.

39-Blum A, Porat R, Rosenschein U, et al. Clinical and inflammatory effects of dietary L-arginine in patients with intractable angina pectoris. Am J Cardiol 1999; 83:1488-1490.

40- Ceremuzynski L, Chamiec T, Herbaczynska-Cedro K. Effect of supplemental oral L-arginine on exercise capacity in patients with stable angina pectoris. Am J Cardiol 1997; 80:331-333.

41- Walker HA, McGing E, Fisher I, et al. Endothelium dependent vasodilation is independent of the plasma L-arginine/ADMA ratio in men with stable angina: lack of effect of oral L-arginine on endothelial function, oxidative stress and exercise performance. J Am Coll Cardiol 2001; 38:499-505.

42- Bednarz B, Wolk R, Chamiec T, et al. Effects of oral L-arginine supplementation on exercise-induced QT dispersion and exercise tolerance in stable angina pectoris. Int J Cardiol 2000; 75:205-210.

43- Egashira K, Hirooka Y, Kuga T, et al. Effects of Larginine supplementation on endothelium-dependent coronary vasodilation in patients with angina pectoris and normal coronary arteriograms. Circulation 1996; 94:130- 134.

44- Bellamy MF, Goodfellow J, Tweddel AC,et al. Syndrome X and Endothelial Dysfunction .Cardiovasc Res. 1998 ;40(2):410-7.

45-Koifman B, Wollman Y, Bogomolny N,et al, Improvement of Cardiac Performance by Intravenous Infusion of L-Arginine in Patients with Moderate Congestive Heart Failure Int J Cardiol. 2002 Dec; 86(2-3):317-23.

 46-Rector TS, Bank A, Mullen KA, et al. Randomized, double-blind, placebo controlled study of supplemental oral L-arginine in patients with heart failure. Circulation 1996; 93:2135-2141.

47- Watanabe G, Tomiyama H, Doba N. Effects of oral administration of L-arginine on renal function in patients with heart failure. J Hypertens 2000; 18:229-234.

48- Yousufuddin M, Flather M, Shamim W, et al. A short course of L-arginine improves exercise capacity and endothelial function in chronic heart failure: a
prospective, randomised, double blind trial. J Am Coll Cardiol 2001:211A.

49-Houghton JL, Toresoff MT, Kuhner PA, et al. African American race predicts improvement in coronary microvascular endothelial function after L-arginine. J Am

31

Coll Cardiol 2001:258A.

50-Sanders PW. Salt-sensitive hypertension: lessons from animal models. Am J Kidney Dis 1996; 28:775-782.

51-Calver A, Collier J, Vallance P. Dilator actions of arginine in human peripheral vasculature. Clin Sci 1991; 81:695- 700.

52- Pezza V, Bernardini F, Pezza E, et al. Study of supplemental oral L-arginine in hypertensives treated with enalapril + hydrochlorothiazide. Am J Hypertens 1998; 11:1267-1270.

53- Siani A, Pagano E, Iacone R, et al. Blood pressure and metabolic changes during dietary L-arginine supplementation in humans. Am J Hypertens 2000; 13:547-551.

54- Maccario M, Oleandri SE, Procopio M, et al. Comparison among the effects of arginine, a nitric oxide precursor, isosorbide dinitrate and molsidomine, two nitric oxide donors, on hormonal secretions and blood pressure in man. J Endocrinol Invest 1997; 20:488-492.

55- Schulze-Neick I, Penny DJ, Rigby ML, et al. L-arginine and substance P reverse the pulmonary endothelial dysfunction caused by congenital heart surgery. Circulation 1999; 100:749-755.

56-Lekakis JP, Papathanassiou S, Papaioannou TG. Oral L-Arginine Improves Endothelial Dysfunction in Patients with Essential Hypertension. J Clin Invest. 1996 15; 97(8):1989-94.

57- Boger RH, Bode-Boger SM, Thiele W, et al. Restoring vascular nitric oxide formation by L-arginine improves the symptoms of intermittent claudication in patients with peripheral arterial occlusive disease. J Am Coll Cardiol 1998; 32:1336- 1344.

58- Roberts JM. Objective evidence of endothelial dysfunction in preeclampsia. Am J Kidney Dis 1999; 33:992-997.

59- Helmbrecht GD, Farhat MY, Lochbaum L, et al. L-arginine reverses the adverse pregnancy changes induced by nitric oxide synthase inhibition in the rat. Am J Obstet Gynecol 1996; 175:800-805.

60- Facchinetti F, Longo M, Piccinini F, et al. L-arginine infusion reduces blood pressure in preeclamptic women through nitric oxide release. J Soc Gynecol Invest 1999; 6:202-207.

61-Palloshi A, Fragasso G, Piatti P,et al. Effect of Oral L-Arginine on Blood Pressure and Symptoms and Endothelial Function in Patients with Systemic Hypertension, Positive Exercise Tests, and Normal Coronary Arteries Cardiovasc Res. 1998 ;40(2):410-7..

5-L-Arginine research in diabetes

Chronic metabolic changes caused by diabetes establish an inflammatory state, accelerate atherogenesis, and increases the risk of cardiovascular fatality [1, 2]. Cellular studies showed that exposure to high glucose induces several metabolic abnormalities that include an increase the activity of the polyol pathway, activation of protein kinase C (PKC), and the generation of reactive oxygen species (ROS) [3]. Of these, activation of the polyol pathway is a key metabolic alteration, which appears to be upstream of PKC and ROS, because inhibition of the aldose reductase (AR), an enzyme that catalyzes the first and the rate-limiting step of the polyol pathway prevents PKC activation and ROS generation [4]. Another major biochemical alteration during diabetes is a decrease in NO bioavailability. Total vascular production of NO is decreased in diabetes[1,2] and gene transfer of endothelial NOS[5] or over-expression of GTP cyclohydrolase I [6], which generates the NOS cofactor , tetrahydrobiopterin, diminishes T1Dinduced endothelial dysfunction. The relationship between NO and the metabolic abnormalities due to AR, PKC, and ROS remains unclear. In vitro, nitroglutathione inhibits AR by inducing S-glutathiolation of the protein4. Thus, NO could potentially regulate biochemical pathways of diabetic injury by inhibiting AR.

Increasing NO production by L-arginine treatment will inhibit AR and the downstream events leading to PKC activation and ROS production. Treatment with L-arginine has been shown before to regulate hyperglycemia and dyslipidemia [7] and inhibit the polyol pathway [8] in diabetic rats.

A study tested the hypothesis that activation of the polyol pathway and protein kinas C (PKC) during diabetes is due to loss of NO showed that after 4 weeks of streptozotocin-induced diabetes, treatment with L-arginine restored NO levels and prevented tissue accumulation of sorbitol in mice, which was accompanied by an increase in glutathiolation of aldose reductase? L-arginine treatment decreased superoxide generation in the aorta, total PKC activity and PKC-βII phosphorylation in the heart, and the plasma levels of triglycerides and soluble ICAM. This study suggested that increasing NO bioavailability by L-arginine corrects the major biochemical abnormalities of diabetes [9].

L-arginine treatment ameliorates vascular inflammation and diabetic dyslipidemia in a murine T1D model. Treatment with L-arginine markedly prevented tissue sorbitol accumulation, ROS generation, and PKC activation - three critical biochemical abnormalities associated with hyperglycemic injury. Taken together, these results support the novel concept that NO is an endogenous negative regulator of the AR-PKC-ROS pathway and therefore a decrease in NO production or availability during diabetes is a key step that is mechanistically linked to the major biochemical effects of high glucose. Increasing NO synthesis by L-arginine could prevent oxidative stress and inflammation by restoring the regulatory axis of NO that controls glucose metabolism via AR and the downstream activation of PKC and ROS production [9].

Changes in NO signaling and production profoundly affect cardiovascular disease [10, 11]. A decrease in NO has been suggested to be the underlying mechanism common to all major CVD risk factors such as high cholesterol, hypertension, and smoking [10, 11]. Diabetes is associated with profound impairment of NO production and signaling. The insulin-resistant state of T2D is associated with marked endothelial dysfunction, which has been variably linked to a decrease in NO synthesis or NO bioavailability [1, 2]. Diabetes has also been associated with increased NOS uncoupling leading to superoxide generation [12].

Endothelium-dependent vascular relaxation is impaired in type 1 and type 2 diabetes mellitus (DM), and endothelial NO deficiency is a likely explanation[13] Diabetes is associated with reduced plasma levels of arginine,[14] and evidence suggests arginine supplementation may be an effective way to improve endothelial function in individuals with diabetes.

An IV bolus of 3-5 g arginine reduced blood pressure and platelet aggregation in patients with type 1 diabetes[15] Low-dose IV arginine improved insulin sensitivity in obese patients and type 2 DM patients as well as in healthy subjects[16]. Arginine may also counteract lipid peroxidation and thereby reduce microangiopathic long-term complications of DM [17]. After one week of oral arginine supplementation (9 g daily), 10 women with type 2 DM showed significant improvement in endothelial function, noted by a 50-percent increase in flow-mediated brachial dilation[18]

A double-blind trial found oral arginine supplementation (3 g three times daily) significantly improved, but did not completely normalize, peripheral and hepatic insulin sensitivity in patients with however, oral arginine (7 g twice daily for six weeks) failed to improve endothelial function type 2 diabetes. In young patients with type 1 DM, however, oral arginine (7 g twice daily for six weeks) failed to improve endothelial function [19, 20].

Reference

1. JA, Beckman; MA, Creager; P, Libby. Diabetes and atherosclerosis: epidemiology, pathophysiology, and management. JAMA 2002; 287:2570–2581.

2. MA, Creager; TF, Luscher; Cosentino, F.; Beckman, JA. Diabetes and vascular disease: pathophysiology, clinical consequences, and medical therapy: Part I. Circulation 2003; 108:1527–1532.

3. Brownlee M. Biochemistry and molecular cell biology of diabetic complications. Nature 2001; 414:813–820.

4. Srivastava SK, Ramana KV, Bhatnagar A. Role of aldose reductase and oxidative damage in diabetes and the consequent potential for therapeutic options. Endocr Rev 2005; 26:380–392.

5. Lund DD, Faraci FM, Miller FJ Jr. Heistad DD. Gene transfer of endothelial nitric oxide synthase improves relaxation of carotid arteries from diabetic rabbits. Circulation 2000; 101:1027–1033.

6. Alp NJ, Mussa S, Khoo J, Cai S, Guzik T, Jefferson A, Goh N, Rockett KA, Channon KM. Tetrahydrobiopterin-dependent preservation of nitric oxide-mediated endothelial function in diabetes by targeted transgenic GTP-cyclohydrolase I over-expression. J Clin Invest 2003; 112:725–735.

7. Mendez JD, Balderas F. Regulation of hyperglycemia and dyslipidemia by exogenous L-arginine in diabetic rats. Biochimie 2001; 83:453–458. [PubMed: 11368855]

8. Chandra D, Jackson EB, Ramana KV, Kelley R, Srivastava SK, Bhatnagar A. Nitric oxide prevents aldose reductase activation and sorbitol accumulation during diabetes. Diabetes 2002; 51:3095–3101.

9-West MB, Ramana KV, Karin Kaiserova K, Srivastava SK, Bhatnagar A. L-arginine prevents metabolic effects of high glucose in diabetic mice. FEBS Lett. 2008; 23; 582(17): 2609–2614.

10. Cooke JP, Dzau VJ. Nitric oxide synthase: role in the genesis of vascular disease. Annu Rev Med 1997; 48:489–509. [PubMed: 9046979]

11. Landmesser U, Hornig B, Drexler H. Endothelial function: a critical determinant in atherosclerosis? Circulation 2004; 109:II27–II33. [PubMed: 15173060]

12. Forstermann U, Munzel T. Endothelial nitric oxide synthase in vascular disease: from marvel to menace. Circulation 2006; 113:1708–1714.

13- Pieper GM. Review of alterations in endothelial nitric oxide production in diabetes. Hypertension 1998; 31:1047–1060.

14-Pieper GM, Siebeneich W, Dondlinger LA. Shortterm oral administration of L-arginine reverses defective endothelium-dependent relaxation and cGMP generation in diabetes. Eur J Pharmacol 1996; 317:317-320.

15- Giugliano D, Marfella R, Verrazzo G, et al. L-arginine for testing endothelium-dependent vascular functions in health and disease. Am J Physiol 1997; 273:E606-E612.

16- Wascher TC, Graier WF, Dittrich P, et al. Effects of low-dose L-arginine on insulin-mediated vasodilatation and insulin sensitivity. Eur J Clin Invest 1997; 27:690-695.

17- Lubec B, Hayn M, Kitzmuller E, et al. L-arginine reduces lipid peroxidation in patients with diabetes mellitus. Free Radic Biol Med 1997; 22:355-357.

18-Regensteiner JG, Popylisen S, Bauer TA, et al. Oral l-arginine and vitamins E and C improve endothelial function in women with type 2 diabetes. Vasc Med 2003; 8:169-175.

19--Piatti PM, Monti LD, Valsecchi G, et al. Long-term oral L-arginine administration improves peripheral and hepatic insulin sensitivity in type 2 diabetic patients. Diabetes Care 2001; 24:875-880.

20- Mullen MJ, Wright D, Donald AE, et al. Atorvastatin but not L-arginine improves endothelial function in type I diabetes mellitus: a double-blind study. J Am Coll Cardiol 2000; 36:410-416.

6-L-arginine and senility

Cardiovascular disease remains the most salient killer in the United States and ageing is a primary risk factor. With advancing age, there is pervasive macrovascular and microvascular dysfunction that manifests as stiffening of central elastic arteries, thickening of intimal–medial layers of the vascular wall, and diminished peripheral conduit artery/resistance artery vasodilatory capacity. Reduced endothelium-dependent vasodilatation has been shown in several prospective studies to be an independent predictor of adverse cardiovascular events.

In the healthy vasculature, nitric oxide (NO) is released from endothelial cells in response to laminar shear stress and causes vasodilatation. In addition to its vasodilatory properties, NO is antiatherogenic: it inhibits platelet aggregation/adhesion, smooth muscle cell proliferation and lipid oxidation.

Vascular and associated ventricular stiffness is one of the hallmarks of the aging cardiovascular system. Both an increase in reactive oxygen species production and a decrease in nitric oxide (NO) bioavailability contribute to the endothelial dysfunction that underlies this vascular stiffness, independent of other age-related vascular pathologies such as atherosclerosis. The activation/upregulation of arginase appears to be an important contributor to age-related endothelial dysfunction by a mechanism that involves substrate (L-arginine) limitation for NO synthase (NOS) 3 and therefore NO synthesis. Not only does this lead to impaired NO production but also it contributes to the enhanced production of reactive oxygen species by NOS. Although arginase abundance is increased in vascular aging models, it appears that posttranslational modification by S-nitrosylation of the enzyme enhances its activity as well. The S-nitrosylation is mediated by the induction of NOS2 in the endothelium. Furthermore, arginase activation contributes to aging-related vascular changes by mechanisms that are not directly related to changes in NO signaling, including polyamine-dependent vascular smooth muscle proliferation and collagen synthesis. Taken together, arginase may represent a target for the modification of age-related vascular and ventricular stiffness contributing to cardiovascular morbidity and mortality.

Depletion of NO has been linked to several pathologies including atherosclerosis and hypertension. Understanding the mechanistic aspects of

NO bioavailability, and loss thereof with ageing, has significant clinical relevance. Synthesizing NO requires a precise admixture of substrate and cofactors. Unidirectional laminar shear stress activates endothelial nitric oxide synthase (eNOS) via phosphorylation. L-Arginine is hydroxyl Ted to *N*-hydroxy-l-arginine and then further oxidized to NO and l-citrulline.NO diffuses into smooth muscle cells, activates guanylate cyclase and induces cyclic GMP-mediated smooth muscle relaxation. Other substrate and cofactors required for this reaction include oxygen, NADPH, flavin, heme and tetrahydrobiopterin (BH4). Altering any one of these variables with ageing may set the reaction awry and attenuate vasodilatation [1-5].

A study examined several potential mechanisms that may contribute to reduced microvascular vasodilatory capacity with ageing [1]. Arterioles from the soleus muscles of young and old rats harvested and exposed to graded increases in intraluminal flow in the absence of changes in intraluminal pressure. Results revealed that arterioles from old rats exhibited a 52% reduction in flow-mediated dilatation (FMD) compared to arterioles from young rats.

Human exercise capacity declines with advancing age and many individuals lose the inclination to participate in regular physical activity. These changes often result in loss of physical fitness and more rapid senescence. A dietary supplement that increases exercise capacity might preserve physical fitness and improve general health and well being in older humans.

Endothelial nitric oxide synthase (eNOS) uses the amino acid L-arginine as a substrate to synthesize nitric oxide (NO). When released from endothelium cells, NO can dilate arteries to increase blood flow [6], help maintain endothelial elasticity [7], prevent platelets from adhering to artery walls [8], mediate erections through smooth muscle relaxation [9], and increase capacity for exercise [10]. In addition, NO can play an integral part in the immune system [11], assist in memory function [12] and sleep regulation [13]. In general, youthful, healthy and athletic individuals have a healthier eNOS system, compared to sedentary, unhealthy and aging individuals [14]. A healthy NO and vascular system facilitates the healthy function of arterioles that mediate oxygen delivery to multiple organs and tissues, including the muscles and kidneys that may impact exercise performance [15].

NO production diminishes in quantity and availability as we age and is associated with an increased prevalence of other cardiovascular risk factors [16].

Hypertension has been shown to promote premature aging of the endothelial system in humans [16]. In individuals with cardiovascular risk factors including hypertension, hypercholesterolemia, smoking, diabetes, obesity, insulin resistance, erectile dysfunction, and metabolic changes associated with aging, supplementation with arginine has been shown to improve NO-dependent endothelial relaxation [17], and improving age-associated endothelial dysfunction [18].

Antioxidants may prevent nitric oxide inactivation by oxygen free radicals. For example, Vitamin C has been shown to improve impaired endothelial vasodilatation in essential hypertensive patients, and effect that can be reversed by the nitric oxide synthase inhibitor NG monomethyl- L-arginine [19]. There is also research indicating that the combination of vitamin C, vitamin E (1.0% to water) and L-arginine works synergistically to enhance nitric oxide production, through nitric oxide synthase gene expression [20]. A study in Atherosclerosis showed Vitamin E (1000 IU/day) improved endothelium health and increased eNOS expression in hypercholesterolemic subjects [21].

References

1-Delp MD, Behnke BJ, Spier SA, Wu G, Muller-Delp JM. Ageing diminishes endothelium-dependent vasodilatation and tetrahydrobiopterin content in rat skeletal muscle arterioles. J Physiol 2008; 586, 1161–1168.

2-DuranteW, Johnson FK, Johnson RA. Arginase: a critical regulator of nitric oxide synthesis and vascular function. Clin Exp Pharmacol Physiol 2007; 34, 906–911.

3-Schmidt TS, Alp NJ. Mechanisms for the role of tetrahydrobiopterin in endothelial function and vascular disease. Clin Sci (Lond) 2007; 113, 47–63.

4-Holowatz LA, Thompson CS, Kenney. Acute ascorbate supplementation alone or combined with arginase inhibition augments reflex cutaneous vasodilatation in aged human skin. Am J Physiol Heart Circ Physiol 2006; 291, H2965–H2970.

5-Eskurza I, Myerburgh LA, Kahn ZD, Seals DR. Tetrahydrobiopterin augments endothelium-dependent dilatation in sedentary but not in habitually exercising older adults. J Physiol 2005; 568, 1057–1065.

6-Wu G, Meininger CJ: Regulation of nitric oxide synthesis by dietary factors. Annu Rev Nutr 2002; 22:61-86.

7- Kinlay S, Creager MA, Fukumoto M, Hikita H, Fang JC, Selwyn AP, Ganz P: Endothelium-derived nitric oxide regulates arterial elasticity in human arteries in vivo. Hypertension 2001; 38(5):1049-53.

8- Preli RB, Klein KP, Herrington DM: Vascular effects of dietary L-arginine supplementation. Atherosclerosis 2002; 162(1):1-15.

9- Mills TM, Pollock DM, Lewis RW, Branam HS, Wingard CJ: Endothelin-1-induced vasoconstriction is inhibited during erection in rats. Am J Physiol Regul Integr Comp Physiol 2001; 281(2):R476-R483.

10- Maxwell AJ, Ho HV, Le CQ, Lin PS, Bernstein D, Cooke JP: L-arginine enhances aerobic exercise capacity in association with augmented nitric oxide production. J Appl Physiol 2001; 90(3):933-8.

11- Marletta MA, Spiering MM: Trace elements and nitric oxide function. J Nutr 2003; 133(5 Suppl 1):1431S-3S.

12 Rickard NS, Ng KT, Gibbs ME: Further support for nitric oxide-dependent memory processing in the day-old chick. Neurobiol Learn Mem 1998; 69(1):79-86.

13-Chen L, Majde JA, Krueger JM: Spontaneous sleep in mice with targeted disruptions of neuronal or inducible nitric oxide synthase genes. Brain Res 2003; 973(2):214-22.

14- Taddei S, Virdis A, Ghiadoni L, Salvetti G, Bernini G, Magagna A, Salvetti A: Age-related reduction of NO availability and oxidative stress in humans. Hypertension 2001; 38(2):274-9.

15- Severs NJ: The cardiac muscle cell. Bioessays 2000; 22(2):188-99.

16- Taddei S, Virdis A, Mattei P, Ghiadoni L, Fasolo CB, Sudano I, Salvetti A: Hypertension causes premature aging of endothelial function in humans. Hypertension 1997; 29(3):736-43.

17- Wu G, Meininger CJ: Arginine nutrition and cardiovascular function. J Nutr 2000; 130(11):2626-9.

18- Chauhan A, More RS, Mullins PA, Taylor G, Petch C, Schofield PM: Aging associated endothelial dysfunction in humans is reversed by Larginine. J Am Coll Cardiol 1996; 28(7):1796-804.

19- Taddei S, Virdis A, Ghiadoni L, Magagna A, Salvetti A: Vitamin C improves endothelium-dependent vasodilation by restoring nitric oxide activity in essential hypertension. Circulation 1998; 97(22):2222-9.

20-de NF, Lerman LO, Ignarro SW, Sica G, Lerman A, Palinski W, Ignarro LJ, Napoli C: Beneficial effects of antioxidants and L-arginine on oxidation sensitive gene expression and endothelial NO synthase activity at sites of disturbed shear stress. Proc Natl Acad Sci USA 2003; 100(3):1420-5.

21- Rodriguez JA, Grau A, Eguinoa E, Nespereira B, Perez-Ilzarbe M, Arias R, Belzunce MS, Paramo JA, Martinez-Caro D: Dietary supplementation with vitamins C and E prevents down regulation of endothelial NOS expression in hypercholesterolemia in vivo and in vitro. Atherosclerosis 2002; 165(1):33-40.

7-L-arginine and athletic performance

The role of nitric oxide in cardiovascular health has been well described in literature. The effect of nitric oxide on exercise performance has been suggested.

During a 5 week progressive strength training program, volunteers were given a supplement containing 1 g arginine and 1 g ornithine, or a placebo, each day. The results suggest that the combination of arginine and ornithine taken in conjunction with a high intensity strength training program can significantly increase muscle strength and lean body mass [1].

Campbell et al [2] observed that arginine and α-ketoglutarate positively influenced 1 RM bench press and Wingate peak power performance in trained adult men. Arginine was also reported to improve peak power significantly in non-athlete men [3]. Conversely, a number of studies have failed to identify any beneficial effect of arginine supplementation.

Liu et al [4] investigated the effect of three day supplementation of 6 gram of arginine on performance in intermittent exercise in well-trained male college judo athletes and found the supplementation had no effect on performance. Similarly, it has been shown that supplementation of arginine aspartate for 14 days prior to marathon run did not affect the subsequent performance in trained runners [5].

Youthful, healthy, athletic individuals generally have a healthier NO system, compared with aging, unhealthy, sedentary individuals [6]. In humans, exercise capacity declines with advancing age and many individuals lose mthe inclination to participate in regular physical activity. In healthy adults, arginine can be synthesized in sufficient quantities to meet most normal physiological demands with the rate of de novo synthesis remaining unaffected by several days of an arginine free diet [7, 8].

L-arginine, vitamin C, and vitamin E promote a healthy cardiovascular system by supporting enhanced NO production [9]. NO formation is further increased by the recycling effect of L-citrulline to L-arginine and the fact that L-citrulline is taken up into cells by a mechanism independent of that for arginine [10].

One study [11] investigated the effects of an L-arginine and antioxidant supplement on exercise performance in elderly male cyclists. The study was a two-arm prospectively randomized double-blinded and placebo-controlled trial. Sixteen male cyclists were randomized to receive either a proprietary supplement (Niteworks®, Herbalife International Inc., and Century City, CA) or a placebo powder. In the control group, there was no change in anaerobic threshold at weeks 1 and 3 compared to baseline. The anaerobic threshold for the supplement groups was significantly higher than that of placebo group at week 1 and week 3.This study suggested that an arginine and antioxidant-containing supplement increase the anaerobic threshold at both week one and week three in elderly cyclists. No effect on VO2 max was observed. This study indicated a potential role of L-arginine and antioxidant supplementation in improving exercise performance in elderly.

In rats, NO stimulates secretion of GH-releasing hormone (GHRH), thereby increasing secretion of GH. GHRH then increases production of NO in somatotroph cells, which subsequently inhibits GH secretion. In humans, arginine stimulates release of GH from the pituitary gland in some populations, but the mechanism is not well understood. Most studies suggest inhibition of somatostatin secretion is responsible for the effect [12].

At high doses (approximately 250 mg/kg body weight), arginine aspartate increased GH secretion,[12] an effect of interest to body builders wishing to take advantage of the anabolic properties of the hormone[13] In a controlled clinical trial, arginine and ornithine (500 mg of each, twice daily, five times per week) produced a significant decrease in body fat when combined with exercise [14]. Acute dosing of arginine (5 g taken 30 minutes before exercise) did not increase GH secretion, and may have impaired release of GH in young adults [15]. Longer-term, low dose supplementation of arginine and ornithine (1 g each, five days per week for five weeks) resulted in higher gains in strength and enhancement of lean body mass, compared with controls receiving vitamin C and calcium[16].

Growth hormone has been observed to be lower in older males than young men. Oral arginine/lysine (3 g each daily) does not increase growth hormone or insulin-like growth factor-I in old men [17].

References

1-Elam RP, Hardin DH, Sutton RA, Hagen L: Effects of arginine and ornithine on strength, lean body mass and urinary hydroxyproline in adult males. J Sports Med Phys Fitness 1989; 29(1):52-6.

2- Campbell B, Roberts M, Kerksick C, Wilborn C, Marcello B, Taylor L, Nassar E, Leutholtz B, Bowden R, Rasmussen C, Greenwood M, Kreider R: Pharmacokinetics, safety, and effects on exercise performance of L-arginine alpha-ketoglutarate in trained adult men. Nutrition 2006; 22(9):872-81.

3- Little JP, Forbes SC, Candow DG, Cornish SM, Chilibeck PD: Creatine, arginine alpha-ketoglutarate, amino acids, and medium-chain triglycerides and endurance and performance. Int J Sport Nutr Exerc Metab 2008; 18(5):493-508.

4- Liu TH, Wu CL, Chiang CW, Lo YW, Tseng HF, Chang CK: No effect of short-term arginine supplementation on nitric oxide production, metabolism and performance in intermittent exercise in athletes. J Nutr Biochem 2009; 20(6):462-8.

5-Colombani PC, Bitzi R, Frey-Rindova P, Frey W, Arnold M, Langhans W, Wenk C: Chronic arginine aspartate supplementation in runners reduces total plasma amino acid level at rest and during a marathon run. Eur J Nutr 1999; 38(6):263-70.

6-Taddei S, Virdis A, Ghiadoni L, Salvetti G, Bernini G, Magagna A, Salvetti A: Age-related reduction of NO availability and oxidative stress in humans. Hypertension 2001; 38(2):274-9.

7-Castillo L, deRojas TC, Chapman TE, Vogt J, Burke JF, Tannenbaum SR, Young VR: Splanchnic metabolism of dietary arginine in relation to nitric oxide synthesis in normal adult man. Proc Natl Acad Sci USA 1993, 90(1):193-7.

8-Castillo L, Ajami A, Branch S, Chapman TE, Yu YM, Burke JF, Young VR: Plasma arginine kinetics in adult man: response to an arginine-free diet. Metabolism 1994; 43(1):114-22.

9-de NF, Lerman LO, Ignarro SW, Sica G, Lerman A, Palinski W, Ignarro LJ, Napoli C: Beneficial effects of antioxidants and L-arginine on oxidationsensitive gene expression and endothelial NO synthase activity at sites of disturbed shear stress. Proc Natl Acad Sci USA 2003; 100(3):1420-5.

10-Wu G, Morris SM Jr: Arginine metabolism: nitric oxide and beyond. *Biochem J* 1998; 336(Pt 1):1-17.

11- Steve Chen, Woosong Kim, Susanne M Henning, Catherine L Carpenter and Zhaoping Li. Arginin and antioxidant supplement on performance in elderly male cyclists: a randomized controlled trial. Journal of the International Society of Sports Nutrition 2010, 7:13.

12- Besset A, Bonardet A, Rondouin G, et al. Increase in sleep related GH and Prl secretion after chronic arginine aspartate administration in man. Acta Endocrinol 1982; 99:18-23.

13- Macintyre JG. Growth hormone and athletes. Sports Med 1987; 4:129-142.

14- Elam RP. Morphological changes in adult males from resistance exercise and amino acid supplementation. J Sports Med Phys Fitness1988; 28:35-39.

15- Marcell TJ, Taaffe DR, Hawkins SA, et al. Oral arginine does not stimulate basal or augment exercise-induced GH secretion in either young or old adults. J Gerontol A Biol Sci Med Sci 1999; 54: M395-M399.

16-Elam RP. Effect of arginine and ornithine on strength, lean body mass and urinary hydroxyproline in adult males. J Sports Nutr 1989; 29:52-56.

17-Corpas E, Blackman MR, Roberson R, et al. Oral arginine-lysine does not increase growth hormone or insulin-like growth factor-I in old men. J Gerontol 1993; 48:M128-M133.

8-L-Arginine and Alzheimer's disease

Alzheimer's disease (AD) is an age-related neurodegenerative disease with an insidious onset. It is characterized by memory impairment and cognitive disturbances that become increasingly more severe with disease progression. It is a debilitating illness causing enormous suffering on the patients and their families, and on society. Approximately 4.5 million Americans are currently affected by AD [1].In the absence of an effective strategies to treat or prevent AD [2], it is expected to affect up to 9 million people by 2040 as the elderly population grows.

The neuropathology of AD is characterized by senile plaques, neurofibrillary tangles (NFT), and, neuronal loss [3-6].The exact causes of AD are still unknown, however studies suggest that the genesis of sporadic AD is associated with atherosclerosis, redox stress, inflammatory processes, and/or abnormal neurotransmission and brain glucose metabolism. Current treatment strategies are limited to altering cholinergic and NMDA neurotransmission and show only modest efficacy. No treatments are currently available to target the underlying mechanism of the disease.

L-arginine is involved in two major metabolic pathways. The nitric oxide synthase (NOS) pathway where L-arginine is converted to NO and L-citruline [10, 11].

There are three isoforms of NOS that are named according to the cell types from which they were first isolated: neuronal NOS (nNOS), inducible NOS (iNOS) and endothelial NOS (eNOS) [10-15]. The synthesis of eNO can be enhanced as a response in a concentration-dependent manner to the increase of extracellular L-arginine concentration [13, 15].This phenomena plays a vital role in the vascular homeostasis associated with L-arginine [16]. The expression of iNOS is induced in inflammatory cell types by cytokine stimulation, and its activity is independent of calcium, and production rate of inducible NO (iNO) is high [17].

L-arginine and NO affect the cardiovascular system as endogenous antiatherogenic molecules that protect the endothelium, modulate vasodilatation, and interact with the vascular wall and circulating blood cells [18-22]. L-arginine and NO can function in the brain as noradrenergic, non-cholinergic neuro-transmitters in learning and memory, synaptic plasticity, and neuroprotection [23, 24]. They can influence the immune system too by playing a key role in regulating inflammatory processes [25] and redox stress. They can also modulate the metabolism of glucose and insulin activity as natural constituents from diets [26] and regulate neurogenesis. L-arginine can possibly affect the AD pathogenesis. Table-8:1 shows the relation between the pathogenesis of sporadic AD and L-arginine.

Pathogenesis of sporadic AD	Effects of L- arginine as a precursor of NO
Atherosclerosis	Anti-atherosclerosis
Redox stress	Modulate Redox stress
Inflammatory processes	Influence the immune system too by playing a key role in regulating inflammatory processes
Abnormal neurotransmission	L-arginine and NO can function as noradrenergic, non-cholinergic neuro-transmitters
Abnormal brain glucose metabolism	Modulate the metabolism of glucose

Table-8:1: The relation between the pathogenesis of sporadic AD and L-arginine.

The second metabolic pathway that involves L-arginine is the arginase pathway where L-arginine is broken down into urea and L-ornithine and genesis of polyamines including putrescine, spermidine, and spermine [27, 28]. Two isoforms of arginase (AI and AII). Depending on the distribution and expression of these isoforms AI and AII might participate in many physiological processes, including inflammation, neurogenesis and apoptosis [29-44].

Polyamines are the major products of L-arginine metabolized by arginase. Ornithine acts as a starting substrate to be converted into putrescine, spermidine and spermine. There are three main polyamines that can be identified with their different lengths of carbon chains [45, 46]. They act as variably functional molecules that are essential for cell regeneration, tissue growth, and development [47-51].

Increasing evidence suggests a strong relationship between AD and atherosclerosis. Some investigators have proposed that AD is a primary neurovascular disease [52-67].Some studies have shown that treatment of atherosclerosis may also benefit AD. Sparks et al suggested that administration of atorvastatin to patients with AD may attenuate cognitive decline and generally slow down the progression of mild-to-moderate AD [68]. That study agrees with others in which statins were used as the treatment for AD [69-71]. Petanceska et al even found that administration of atorvastatin can significantly reduce Aβ amyloid deposition in an animal model [72]. Table (8:2) shows the relationship between AD and L-arginine.

Arginine (1.6 g/day) in 16 elderly patients with senile dementia reduced lipid peroxidation and increased cognitive function [110].

Table – 8:2: The relationship between AD and L-arginine
1-L-Arginine affects AD by anti-atherosclerosis chronic oral supplementation with L-arginine may block the progression of atherosclerotic plaques via restoration of NOS substrate availability, decrease of vascular stress and antihypertensive effect [73-79].
2- L-arginine affects AD by influencing oxidative stress [80-88] L-arginine as a precursor of NO can protect it through exerting its anti-oxidant functions. If lacking L-arginine and NO, the brain would have an increase of superoxide anion formation. Administration of L-arginine may be associated with the antiradical and antioxidant effects of NO, inhibiting the effects of inositol-1, 2, 5-triphophates, and inhibiting the accumulation of leukocytes in the reperfused tissue [89, 90, 91].
3- The effects of L-arginine on by via influencing inflammation L-arginine is a protective molecule, modulating oxidant-mediated neuro-inflammation by the production of NO [92]. L-arginine and NO, as modulators, may play a role in AD by influencing inflammatory processes. Regulating the level and the metabolic pathway of L-arginine and selectively producing different isoforms of NO may produce therapeutic effects.
4-L-Arginine affects through production of the neurotransmitter NO NO, as a transmitter, modulated synaptic efficacy at the neuromuscular junction. They also demonstrated that NO regulates transmitter release and adenosine-induced depression via a cGMP-dependent mechanism which occurs after $Ca2+$ entry [93-99].
5-Effect of L-arginine on neurogenesis Metabolism of L-arginine in the second pathway lead to the genesis of polyamines including putrescine, spermidine, and spermine. Polyamines, products of L-arginine through the arginase pathway, have their specific functions in neurogenesis. Depletion of polyamines during nervous system development will lead to a deficiency of neuronal morphogenesis [100-109].

References

1-Evans DA, Funkenstein HH, Altert MS, Scherr PA, Cook NR, Chown MJ, Herbert LE, Hennekens CH and Taylor JO. Prevalence of Alzheimer's disease in a community population of older persons. JAMA 1989; 262: 2551-2556.

2-U.S. Government Printing Office, Washington, DC. National Institute on Aging (1995) Progress Report on Alzheimer's disease. NIH Publication 1995; 95-3994.

3-Selkoe DJ. Alzheimer's disease: genes, proteins, and therapy. Physiol Rev 2001; 81: 741-766.

4-Yaari R and Corey-Bloom J. Alzheimer's disease. Semin Neurol 2007; 27:32-41.

5-oedert M and Spillantini MG. A century of Alzheimer's disease. Science 2006; 314:777-781.

6-Mattson MP. Pathways towards and away from Alzheimer's disease. Nature 2004; 430: 631-639.

7-Rose WC. The nutritional significance of the amino acids. Physiol Rev 1938; 18:109-136.

8-Hedin SG. Eine methode das lysin zu isolieren, nebst einigen Bemerkungen uber das lysatinin. Z Physiol Chem 1895; 21:297-305.

9-Schulze E and Steiger E. Uber das Arginin. Z Physiol Chem 1886; 11:43-65.

10-Moncada S and Higgs A. The L-arginine-nitric oxide pathway. N Engl J Med 1993; 329: 2002-2012.

11-Sidney M and Morris Jr. Regulation of enzymes of the urea cycle and arginine metabolism. Annu Rev Nutr 2002; 22: 87-105.

12-Forstermann U, Closs EI, Pollock JS, Nakane M, Schwarz P, Gath I and Kleinert H. Nitric oxide synthase isozymes. Characterization, purification, molecular cloning, and functions. Hypertension 1994; 23:1121-1131.

13-Boger RH. The pharmacodynamics of L-arginine. J Nutr 2007; 137(Suppl 2):1650S-1655S.

14-Malinski T. Nitric oxide and nitroxidative stress in Alzheimer's disease. J Alzheimers Dis 2007; 11:207-218.

15-Loscalzo J. What we know and don't know about L-arginine and NO. Circulation 2000; 101:2126-2129.

16-Siasos G, Tousoulis D, Antoniades C, Stefanadi E and Stefanadis C. L-Arginine, the substrate for NO synthesis: An alternative treatment for premature atherosclerosis? Int J Cardiol 2007; 116:300-308.

17-Forstermann U, Boissel JP and Kleinert H. Expressional control of the constitutive isoforms of nitric oxide synthase (NOS I and NOS III). FASEB J 1998; 12:773-790.

18-Boger RH, Bode-Boger SM and Frolich JC. The L-arginine – nitric oxide pathway: Role in atherosclerosis and therapeutic implications. Atherosclerosis 1996; 127:1-11.

19-Cooke JP and Dzau VJ. Nitric oxide synthase: role in the genesis of vascular disease. Annu Rev Med 1997; 48:489-509.

20-Cooke JP. The pivotal role of nitric oxide for vascular health. Can J Cardiol 2004; 20(Suppl B):7B-15B.

21-Li XA, Everson W and Smart EJ. Nitric oxide, caveolae, and vascular pathology. Cardiovasc Toxicol 2006; 6:1-13.

22-Napoli C, de Nigris F, Williams-Ignarro S, Pignalosa O, Sica V and Ignarro LJ. Nitric oxide and atherosclerosis: An update. Nitric Oxide 2006; 15:265-279.

23-Bohme GA, Bon C, Lemaire M, Reibaud M, Piot O, Stutzmann JM, Doble A and Blanchard JC. Altered synaptic plasticity and memory formation in nitric oxide synthase inhibitor-treated rats. Proc Natl Acad Sci USA 1993; 90: 9191-9194.

24-Paakkari I and Lindsberg P. Nitric oxide in the central nervous system. Ann Med 1995; 27: 369-377.

25-Potenza MA, Nacci C and Mitolo-Chieppa D. Immunoregulatory effects of L-arginine and therapeutically implications. Curr Drug Targets Immune Endocr Metabol Disord 2001; 1:67-77.

26-Jobgena WS, Friedb SK, Fuc WJ, Meiningerd CJ and Wu G. Regulatory role for the arginine–nitric oxide pathway in metabolism of energy substrates. J Nutr Biochem 2006; 17:571-588.

27-Wu G and Morris SM Jr. Arginine metabolism: nitric oxide and beyond. Biochem J 1998; 336: 1-17.

28-Iyer R, Jenkinson CP, Vockley JG, Kern RM, Grody WW and Cederbaum S. The human arginases and arginase deficiency. J Inherit Metab Dis 1998; 1:86-100.

29-Sidney M and Morris SM Jr. Regulation of enzymes of the urea cycle and arginine metabolism. Annu Rev Nutr 2002; 22:87-105.

30-Glass RD and Knox WE. Arginase isozymes of rat mammary gland, liver and other issues. J Biol Chem 1973; 248:5785-5789.

31-Kaysen GA and Strecker HJ. Purification and properties of arginase of rat kidney. Biochem J 1973; 133:779-788.

32-Spector EB, Rice SCH and Cederbaum SD. Immunologic studies of arginase in tissues of normal human adults and arginase-deficient patients. Pediatr Res 1983; 17:941-944.

33-Grody WW, Argyle C, Kern RM, Dizikes GJ, Spector EB, Strickland AD, Klein D and Cederbaum SD. Differential expression of the two human arginase genes in hyperargininemia: enzymatic pathologic and molecular analysis. J Clin Invest 1989; 83: 602-609.

34-Lange PS, Langley B, Lu P and Ratan RR. Novel roles for arginase in cell survival, regeneration, and translation in the central nervous system. J Nutr 2004; 134(Suppl): 2812S-2817S.

35-Becker-Catania SG, Gregory TL, Yang Y, Gau CL, de Vellis J, Cederbaum SD and Iyer RK. Loss of arginase I results in increased proliferation of neural stem cells. J Neurosci Res 2006; 84:735-746.

36-Boeshore KI, Schreiber RC, Vaccariello SA, Sachs HH, Salazer R, Lee J, Ratan RR, Leahy P and Zigmond RE. Novel changes in gene expression following axotomy of a sympathetic ganglion: a microarray analysis. J Neurobiol 2004; 59:216-235.

37-Yu H, Iyer RK, Kern RT, Rodriquez WI, Grody WW and Cederbaum SD. Expression of arginase isozymes in mouse brain. J Neurosci Res 2001; 66:406-422.

38-Yu H, Iyer, RK, Yoo PK, Kern RM, Grody WW and Cederbaum SD. Arginase expression in mouse embryonic development. Mech 2002; 115:151-155.

39-Terheggen HF, Schwenk A, Lowenthal A, van Sande M and Colombo JPZ. Hyperargininemia with arginase deficiency, a new familial metabolic disease: clinical aspects. Kinderheilk 1970; 107:298-312.

40-Scaglia F and Lee B. Clinical, biochemical, and molecular spectrum of hyperargininemia due to arginase I deficiency. Am J Med Genet C Semin Med Cenet 2006; 142:113-120.

41-Cederbaum SD, Shaw KNF, Spector EB, Verity MA, Snodgrass PJ and Sugarman GI. Hyperargininemia due to arginase deficiency. Pediatr Res 1979; 13:827-833.

42-Spector EB, Rice SCH and Cederbaum SD. Immunologic studies of arginase in tissues of normal human adults and arginase-deficient patients. Pediatr Res 1983; 17:941-944.

43-Grody WW, Kern RM, Klein D, Dodson AE, Wissman PB, Barsky SH and Cederbaum SD. Arginase deficiency manifesting delayed clinical sequelae and induction of a kidney arginase isozyme. Hum Genet 1993; 91:1-5.

44-Spector EB, Jenkinson CP, Grigor MR, Kern RM and Cederbaum SD. Subcellular location and differential antibody specificity of arginase in tissue culture and whole animals. Int J Dev Neurosci 1994; 12:337-342.

45-Tabor CW and Tabor H. Polyamines. Annu Rev Biochem 1984; 53:749-790.

46-Morgan DML. Polyamines. Essays Biochem 1987; 46: 82-115.

47-Cai D, Deng K, Mellado W, Lee J, Ratan RR and Filbin MT. Arginase I and polyamines act downstream from cyclic AMP in overcoming inhibition of axonal growth MAG and myelin in vitro. Neuron 2002; 35:711-719.

48- Nishioka K. Introduction to polyamines. In: Nishioka K (ed) Polyamines in Cancer: Basic Mechanisms and Clinical Approaches. New York Springer, 1996; pp 1-5.

49- Schipper RG, Penning LC and Verhofstad AA. Involvement of polyamines in apoptosis. Facts and controversies: effectors or protectors? Cancer Biol 2000; 10:55-68.

50- Thomas T and Thomas TJ. Polyamines in cell growth and cell death: molecular mechanisms and therapeutic applications. Cell Mol Life Sci 2001; 58:244-258.

51- Auvinen M, Jarvinen K, Hotti A, Okkeri J, Laitinen J, Janne OA, Coffino P, Bergman M, Andersson LC, Alitalo K and Holtta E. Transcriptional regulation of the ornithine decarboxylase gene by c-Myc/Max/Mad network and retinoblastoma protein interacting with c-Myc. Int J Biochem Cell Biol 2003;35:496-521.

52- de la Torre JC. Alzheimer disease as a vascular disorder: Nosological evidence. Stroke 2002; 33:1152-1162.

53- Rhodin JA and Thomas T. A vascular connection to Alzheimer's disease. Microcirculation 2001, 8:207-220.

54- Breteler MM, Bots ML, Ott A and Hofman A. Risk factors for vascular disease and dementia. Haemostasis 1998; 28:167-173.

55- Roher AE, Esh C, Kokjohn TA, Kalbak W, Luhers DC, Seward JD, Sue LI and Beach TG. Circle of Willis atherosclerosis is a risk factor for sporadic Alzheimer's disease. Arterioscler Thromb Vasc Biol 2003; 23:2055-2062.

56- Breteler MM. Vascular involvement in cognitive decline and dementia: epidemiologic evidence from the Rotterdam Study and the Rotterdam Scan Study. Ann N Y Acad Sci 2000; 903:457-465.

57- Breteler MM. Vascular risk factors for Alzheimer's disease: an epidemiological study. Neurobiol Aging 2000; 21:153-160.

58- Ott A, Stolk RP, Hofman A, van Harskamp F, Grobbee DE and Breteler MM. Association of diabetes mellitus and dementia: the Rotterdam Study. Diabetologia 1996; 39: 1392-1397.

59- Ott A, Slooter AJ, Hofman A, van Harskamp F and Witteman JC. Smoking and risk of dementia and Alzheimer's disease in a population-based cohort study: the Rotterdam Study. Lancet 1998; 351:1840-1843.

60- Van Duijn CM, Havekes LM, van Broeckhoven C, de Knijff P and Hofman A. Apolipoproyein E genotype and association between smoking and early onset Alzheimer's disease. Br Med J 1995; 310:627-631.

61- Graves AB, van Duijn CM, Chandra V, Fratiglioni L, Heyman A, Jorm AF, Kokmen E, Kondo K, Mortimer JA, Rocca WA, Shalat S, Soininen H and Hofman A. Alcohol and tobacco consumption as risk factors for Alzheimer's disease: a collaborative re-analysis of case-controlled studies. Int J Epidemiol 1991; 20:S48-S57.

62- Roher AE, Esh C, Kokjohn TA, Kalback W, Luehrs DC, Seward JD, Sue LI and Beach TG.Circle of willis atherosclerosis is a risk factor for sporadic Alzheimer's disease. Thromb Vasc Biol 2003; 23; 2055-2062.

63- Beach TG, Wilson JR, Sue LI, Newell A, Poston M, Cisneros R, Pandya Y, Esh C, Connor DJ, Sabbagh M, Walker DG and Roher AE. Circle of Willis atherosclerosis: association with Alzheimer's disease, neuritic plaques and neurofibrillary tangles. Acta Neuropathol (Berl) 2007; 113:13-21.

64- Hirao K, Ohnishi T, Hirata Y, Yamashita F, Mori T, Moriguchi Y, Matsuda H, Nemoto K, Imabayashi E, Yamada M, Iwamoto T, Arima K and Asada T. The prediction of rapid conversion to Alzheimer's disease in mild cognitive impairment using regional cerebral blood flow SPECT. Neuroimage 2005; 28: 1014-1021.

65- Johnson KA, Jones K, Holman BL, Becker J, Spiers PA, Satlin A and Albert MS. Preclinical prediction of Alzheimer's disease using SPECT. Neurology 1998; 50:1563-1571.

66- Johnson KA and Albert MS. Perfusion abnormalities in prodromal Alzheimer's disease. Neurobiol Aging 2000; 21:289-292.

67- Matsuda H, Mizumura S, Nagao T, Ota T, Iizuka T, Nemoto K, Kimura M, Tateno A, Ishiwata A, Kuji I, Arai H and Homma A. An easy Z-score imaging system for discrimination between very early Alzheimer's disease and controls using brain perfusion SPECT in a multicentre study. Nucl Med Commun 2007; 28:199-205.

68- Sparks DL, Sabbagh M, Connor D, Soares H, Lopez J, Stankovic G, Johnson-Traver S, Ziolkowski C and Browne P. Statin therapy in Alzheimer's disease. Acta Neurol Scand Suppl 2006; 185:78-86.

69- Zamrini E, McGwin G and Roseman JM. Association between statin use and Alzheimer's disease. Neuroepidemiology 2004; 23:94-98.

70- Miida T, Takahashi A, Tanabe N and Ikeuchi T. Can statin therapy really reduce the risk of Alzheimer's disease and slow its progression? Curr Opin Lipidol 2005; 16:619-623.

71- Zigman WB, Schupf N, Jenkins EC, Urv TK, Tycko B and Silverman W. Cholesterol level, statin use and Alzheimer's disease in adults with Down syndrome. Neurosci Lett 2007; 18; 416:279-284.

72- Petanceska SS, DeRosa S, Olm V, Diaz N, Sharma A, Thomas-Bryant T, Duff K, Pappolla M and Refolo LM. Statin therapy for Alzheimer's disease: will it work? J Mol Neurosci 2002; 19:155-161.

73- Cooke JP and Creager A. Endothelial dysfunction in hypercholesterolemia is corrected by L-arginine. Basic Res Cardiol 1991; 86(Suppl 2):173-181.

74- Dhawan V, Handu SS, Nain CK and Ganguly NK. Chronic L-arginine supplementation improves endothelial cell vasoactive functions in hypercholesterolemic and atherosclerotic monkeys. Mol Cell Biochem 2005; 269:1-11.

75- Jeremy RW, McCarron H and Sullivan D. Effects of dietary L-arginine on atherosclerosis and endothelium-dependent vasodilation in the hypercholesterolemic rabbit. Response according to treatment duration, anatomic site and sex. Circulation 1996; 94:498-506.

76- Verreault R, Kaltenbach G and Berthel M. Hypertension and Alzheimer's disease. Presse Med 2005; 34:809-812.

77- Skoog I and Gustafson D. Update on hypertension and Alzheimer's disease. Neurol Res 2006; 28:605-611.

78- Siani A, Pagano E, Iacone R, Iacoviello L, Scopacasa F and Strazzullo P. Blood pressure and metabolic changes during dietary L-arginine supplementation in humans. Am J Hypertens 2000; 13:547-551.

79- Rector TS, Bank AJ, Mullen KA, Tschumperlin LK, Sih R, Pillai K and Kubo SH. Randomized, double-blind, placebo-controlled study of supplemental oral L-arginine in patients with heart failure. Circulation 1996; 93:2135-2141.

80- Halliwell B. Protection against tissue damage in vivo by desferrioxamine: what is its mechanism of action? Free Radic Biol Med 1989; 7:645-651.

81- Smith MA, Perry G, Richey PL, Sayre LM, Anderson VE, Beal MF and Kowall N. Oxidative damage in Alzheimer's. Nature 1996; 382:120-121.

82- Sayre LM, Zelasko DA, Harris PL, Perry G, Salomon RG and Smith MA. 4-Hydroxynonenal-derived advanced lipid peroxidation end products are increased in Alzheimer's disease. J Neurochem 1997; 68: 2092-2097.

83- Smith MA. Alzheimer disease. Int Rev Neurobiol 1998; 42:1-54.

84- Nunomura A, Perry G, Pappolla MA, Friedland RP, Hirai K, Chiba S and Smith MA. Neuronal oxidative stress precedes amyloid-beta deposition in Down syndrome. J Neuropathol Exp Neurol 2000; 59:1011-1017.

85- Nunomura A, Perry G, Aliev G, Hirai K, Takeda A, Balraj EK, Jones PK, Ghanbari H, Wataya T, Shimohama S, Chiba S, Atwood CS, Petersen RB and Smith MA. Oxidative damage is the earliest event in Alzheimer disease. J Neuropathol Exp Neurol 2001; 60:759-767.

86- Zhu X, Raina AK, Lee HG, Casadesus G, Smith MA and Perry G. Oxidative stress signaling in Alzheimer's disease. Brain Res 2004; 1000: 32-39.

87-Perry G, Taddeo MA, Nunomura A, Zhu X, Zenteno-Savin T, Drew KL, Shimohama S, Avila J, Castellani RJ and Smith MA. Comparative biology and pathology of oxidative stress in Alzheimer and other neurodegenerative diseases: beyond damage and response. Comp Biochem Physiol C Toxicol Pharmacol 2002; 133:507-513.

88-Reddy PH. Amyloid precursor protein-mediated free radicals and oxidative damage: implications for the development and progression of Alzheimer's disease. J Neurochem 2006; 96:1-13.

89-Miliutina NP, Ananian AA and Shugalei VS. Antiradical and antioxidant effect of arginine and its action on lipid peroxidation in hypoxia. Biull Eksp Biol Med 1990; 110:433-435.

90-Maksimovich NE and Maslakov DA. The amino acid L-arginine and the potential for its use in clinical practice. Zdravookhranenie 2003; 5:35-37.

91- Sedlak J and Lindsay RH. Estimation of total, protein-bound, and nonprotein sulfhydryl groups in tissue with Ellman's reagent. Anal Biochem 1968; 25:192-205.

92- Scott GS and Bolton C. L-arginine modifies free radical production and the development of experimental allergic encephalomyelitis. Inflamm Res 2000; 49:720-726.

93-Garthwaite J, Charles SL and Chess-Williams R. Endothelium-derived relaxing factor release on activation of NMDA receptors suggests role as intercellular messenger in the brain. Nature 1988; 336:385-388.

94- Dawson TM, Dawson VL and Snyder SH. A novel neuronal messenger molecule in brain: the free radical, nitric oxide. Ann Neurol 1992; 32:297-311.

95-Bon CL and Garthwaite J. On the role of nitric oxide in hippocampal long-term potentiation. J Neurosci 2003; 23:1941-1948.

96-Sanders KM and Ward SM. Nitric oxide as a mediator of nonadrenergic noncholinergic neurotransmission. Am J Physiol 1992; 262: 379-392.

97- Garthwaite J and Boulton CL. Nitric oxide signaling in the central nervous system. Annu Rev Physiol 1995; 57:683-706.

98-Thomas S and Robitaille R. Differential frequency-dependent regulation of transmitter release by endogenous nitric oxide at the amphibian neuromuscular synapse. J Neurosci 2001; 21:1087-1095.

99- Nickels TJ, Reed GW, Drummond JT, Blevins DE, Lutz MC and Wilson DF. Does nitric oxide modulate transmitter release at the mammalian neuromuscular junction? Clin Exp Pharmacol Physiol 2007; 34:318-326.

100- Schweitzer L, Robbins AJ and Slotkin TA. Dendritic development of Purkinje and granule cells in the cerebellar cortex of rats treated postnatally with alpha-difluoromethylornithine. J Neuropathol Exp Neurol 1989; 48:11-22.

101-Harada J and Sugimoto M. Polyamines cerebellar granule neurons. Brain Res 1997; 753:251-259.

102- Chu PJ, Saito H and Abe K. Polyamines promote regeneration of injured axons of cultured rat hippocampal neurons. Brain Res 1995; 673:233-241.

103- Malaterre J, Strambi C, Aouane A, Strambi A, Rougon G and Cayre M. A novel role for polyamines in adult neurogenesis in rodent brain. Eur J Neurosci 2004; 20:317-330.

104-Cayre M, Malaterre M, Strambi C, Charpin P, Ternaux J-P and Strambi A. Short- and long-chain natural polyamines play specific roles in adult cricket neuroblast proliferation and neuron Differentiation in Vitro. J Neurobiol 2001; 48:315-324.

105Thomas T and Thomas TJ. Regulation of cyclin B1 by estradiol and polyamines in MCF-7 breast cancer cells. Cancer Res 1994; 54: 1077-1084.

106-Thomas T, Gallo MA, Klinge CM and Thomas TJ. Polyamine-mediated conformational perturbations in DNA alter the binding of estrogen receptor to poly (dG-m5dC) z poly (dG-m5dC) and a plasmid containing the estrogen response element. J Steroid Biochem Mol Biol 1995; 54:89-99.

107-Kaminska B, Kaczmarek L and Grzelakowska-Sztabert B. Inhibitors of polyamine biosynthesis affect the expression of genes encoding cytoskeletal proteins. FEBS Lett 1992; 304:198-200.

108- Filhol O, Loue-Mackenbach P, Cochet C and Chambaz EM. Casein kinase II and polyamines may interact in the response of adrenocortical cells to their trophic hormone. Biochem Biophys Res Comm 1991; 180:623-630.

109- Ulloa L, Diaz-Nido J and Avila J. Depletion of casein kinase II by antisense oligonucleotide prevents neuritogenesis in neuroblastoma cells. EMBO J 1993; 12:1633-1640.

110-Ohtsuka Y, Nakaya J. Effect of oral administration of L-arginine on senile dementia. Am J Med 2000; 108:439.

9-L-Arginine in chronic kidney disease

Nitric oxide (NO) production is reduced in renal disease, partially due to decreased endothelial NO production. Endothelial dysfunction occurs in CKD (characterized by blunted release of endothelial NO) and earlier in ESRD, even during early stages of disease. NO deficiency contributes to cardiovascular events and progression of kidney damage [1-11]. Chronic inhibition of NOS in otherwise normal animals produces hypertension and focal segmental glomerulosclerosis, the hallmark of progressive CKD [12]. It therefore seems likely that the NO deficiency associated with CKD contributes to progression of kidney damage and eventual development of ESRD.

L-Arginine is a 'semi-essential' amino acid. During periods of high demand, such as maturational growth or following injury, dietary ingestion of L-arginine is necessary. By contrast, endogenous arginine production satisfies unstressed adults Endogenous arginine synthesis is concentrated in the liver (where arginine is rapidly hydrolyzed in the urea cycle) and kidney cortex. Most L-arginine produced in the kidney is released into the blood and distributed throughout the body [13, 14].

There are two possible causes of NO deficiency are substrate (L-arginine) limitation and increased levels of circulating endogenous inhibitors of NO synthase (particularly asymmetric dimethylarginine [ADMA]).

Decreased L-arginine availability in chronic kidney disease (CKD) is due to disturbed renal biosynthesis of this amino acid. Inhibition of transport of L-arginine into endothelial cells and shunting of L-arginine into other metabolic pathways (e.g. those involving arginase) might also decrease availability. Elevated plasma and tissue levels of ADMA in CKD are functions of both reduced renal excretion and reduced catabolism by dimethylarginine dimethylaminohydrolase (DDAH). DDAH might be associated with loss-of function polymorphisms of a DDAH gene, functional inhibition of the enzyme by oxidative stress in CKD and end-stage renal disease, or both.

The kidney contains arginase II, and endothelial cells contain both isoforms. Arginases compete with NOS for L-arginine and can thereby limit NO production; indeed, NO production is attenuated in endothelial cells overexpressing arginase I and arginase II.55 Vascular arginase activity is increased in experimental hypertension and inhibition of arginase restores endothelial NO synthesis. Arginase I and arginase II are upregulated in the vasculature of Dahl salt-sensitive rats maintained on a high-salt diet; similar findings have been reported for spontaneously hypertensive rats. The inhibition of arginase protects the kidney from structural damage in the 5/6 renal mass ablation/infarction model of CKD, indicating that arginase inhibition might be a novel therapeutic approach to slowing progressive renal damage [15, 19].

Several endogenous NOS inhibitors have been identified, including methylguanidines and methylated arginines such as N-monomethyl-L-arginine (L-NMMA) and ADMA. ADMA is a potent competitive inhibitor of NOS. Acute infusion of ADMA in healthy humans causes renal and systemic vasoconstriction, and causes heart rate and cardiac output to drop. Chronically elevated levels of endogenous ADMA have been implicated in endothelial dysfunction, and increased cardiovascular morbidity and mortality, in many diseases including atherosclerosis, heart failure, hypertension, diabetes and ESRD. Plasma ADMA, which has spilled out of the cells in which it was produced, is the measured variable. In vascular endothelium, ADMA levels are approximately 10-fold those of plasma, and concentrations are extremely high in the kidney and spleen. It is the local intracellular level of ADMA that regulates NOS activity and this probably varies greatly between organs [8, 20-26].

Understanding the role of L-arginine in CKD provide the basis for the rationale for novel therapies, including supplementation of dietary L-arginine or its precursor L-citrulline, inhibition of non-NO producing pathways of L-arginine utilization, or both.

L-arginine, a substrate for nitric oxide synthases, when administered alone acutely by intravenous infusion (500 mg kg-1 over 30 min) decreases blood pressure in both normotensive and hypertensive subjects [27]. L-arginine administered orally at 9–18 g/ day for14 days to hypertensive hemodialysis patients or chronically for 2 months to hypertensive kidney transplant patients resulted in a decrease in both systolic blood pressure and diastolic blood pressure [28].

In patients with ESRD, plasma ADMA predicts carotid artery intima–media thickness and the rate of worsening of carotid atherosclerosis,[29] as well as the extent of left ventricular hypertrophy and dysfunction[30] Increased ADMA levels have been reported in a variety of conditions associated with cardiovascular risk—including essential hypertension[31] and peripheral arterial occlusive disease[8,32]even when renal function is normal. In healthy, male nonsmokers, elevated plasma ADMA concentration is associated with a fourfold increased risk of acute coronary events [33].

Because an increase in ADMA has emerged as a major independent risk factor in end-stage renal disease (and probably also in CKD), lowering ADMA concentration is a major therapeutic goal; interventions that enhance the activity of the ADMA-hydrolyzing enzyme DDAH are under investigation.

References

1- Baylis C, Vallance P. Measurement of nitrite and nitrate (NOx) levels in plasma and urine; what does this measure tell us about the activity of the endogenous nitric oxide. Current Opinion Nephrol Hypertens 1998; 7:1–4.

2- Schmidt R, et al. Indices of activity of the nitric oxide system in patients on hemodialysis. Am J Kidney Dis 1999; 34:228–234.

3-Schmidt RJ, et al. Nitric oxide production is low in end stage renal disease patients on peritoneal dialysis. Am J Physiol 1999; 276:F794–F797.

4- Schmidt RJ, Baylis C. Total nitric oxide production is low in patients with chronic renal disease. Kidney Int 2000; 58:1261–1266.

5- Blum M, et al. Low nitric oxide production in patients with chronic renal failure. Nephron 1998; 79:265–268.

6- Wever R, et al. Nitric oxide production is reduced in patients with chronic renal failure. Arterioscler Thromb Vasc Biol 1999; 19:1168–1172.

7- Thambyrajah J, et al. Abnormalities of endothelial function in patients with predialysis renal failure. Heart 2000; 83:205–209.

8- Böger RH, Zoccali C. ADMA: a novel risk factor that explains excess cardiovascular event rate in patients with end-stage renal disease. Atherosclerosis 2003 ;(Suppl 4):S23–S28.

9- Foley RN, et al. Clinical epidemiology of cardiovascular disease in chronic renal disease. Am J Kidney Dis 1998; 32(Suppl 3):S112–S119.

10- Baigent C, et al. Premature cardiovascular disease in chronic renal failure. Lancet 2000; 356:147–152.

11- Landray MJ, et al. Inflammation, endothelial dysfunction, and platelet activation in patients with chronic kidney disease: the chronic renal impairment in Birmingham (CRIB) study. Am J Kidney Dis 2004; 43:244–253.

12- Zatz R, Baylis C. Chronic nitric oxide inhibition model six years on. Hypertension 1998; 32:958–964.

13-Wu G, Morris SM. Arginine metabolism: nitric oxide and beyond. Biochem J 1998; 33:1–17.

14- Brosnan ME, Brosnan JT. Renal arginine metabolism. J Nutr 2004; 134(Suppl 10):S2791–S2795.

15- Li H, et al. Regulatory role of arginase I and II in nitric oxide, polyamine, and proline syntheses in endothelial cells. Am J Physiol Endocrinol Metab 2001; 280:E75–E82.

16- Sabbatini M, et al. Arginase inhibition slows the progression of renal failure in rats and renal ablation. Am J Physiol Renal Physiol 2003; 284:F680–F687.

17-Zhang C, et al. Upregulation of vascular arginase in hypertension decreases nitric oxide mediated dilation of coronary arterioles. Hypertension 2004; 44:935–943.

18- Johnson FK, et al. Arginase inhibition restores arteriolar endothelial function in Dahl rats with salt induced hypertension. Am J Physiol Regul Integr Comp Physiol 2005; 288:1057–1062.

19- Demougeot C, et al. Arginase inhibition reduces endothelial dysfunction and blood pressure rising in SHR. J Hypertension 2005;23:971–978.

20- McAllister RJ, et al. Effects of guanidino and uremic compounds on nitric oxide pathways. Kidney
Int 1994; 45:737–742.

21- Leiper J, Vallance P. Biological significance of endogenous methylarginines that inhibit nitric oxide synthases. Cardiovasc Res 1999; 43:542–548.

22- Achan V, et al. Asymmetric dimethylarginine causes hypertension and cardiac dysfunction in humans and is actively metabolized by dimethylarginine dimethylaminohydrolase. Arterioscler Thromb Vasc Biol 2003; 23:1455–1459.

23-Kielstein JT, et al. Cardiovascular effects of systemic nitric oxide synthase inhibition with asymmetric dimethylarginine in humans. Circulation 2004; 109:172–177.

24- Vallance P, Leiper J. ADMA and kidney disease—marker or mediator. J Am Soc Nephrol 2005; 16:2254–2256. [PubMed: 15987745]

25- Scalera F, et al. Erythropoietin increases asymmetric dimethylarginine in endothelial cells: role of dimethylarginine dimethylaminohydrolase. J Am Soc Nephrol 2005; 16:892–898.

26- Ueno S, et al. Distribution of free methylarginines in rat tissues and in the bovine brain. J Neurochem 1992; 59:2012–2016.

27-Higashi Y, Oshima T, Ozono R, Matsuura H, Kambe M, Kajiyama G. Effect of l arginine infusion on systemic and renal hemodynamics in hypertensive patients. Am J Hypertension 1999; 12 (Part 1): t-15.

28- Kelly BS, Alexander JW, Dreyer D, Greenberg NA, Erickson A, Whiting JF, et al. Oral arginine improves blood pressure in renal transplant and hemodialysis patients. J Parenteral Enteral Nutrition 2001; 25: 194–202.

29- Zoccali C, et al. Asymmetric dimethylarginine, C–reactive protein, and carotid intima–media thickness in end-stage renal disease. J Am Soc Nephrol 2002; 13:490–496.

30- Zoccali C, et al. Left ventricular hypertrophy, cardiac remodeling and asymmetric dimethylarginine (ADMA) in hemodialysis patients. Kidney Int 2002; 62:339–345.

31- Kielstein JT, et al. Asymmetric dimethylarginine plasma concentrations differ in patients with end stage renal disease: relationship to treatment method and atherosclerotic disease. J Am Soc Nephrol 1999; 10:594–600.

32- Surdacki A, et al. Reduced urinary excretion of nitric oxide metabolites and increased plasma levels of asymmetric dimethylarginine in men with essential hypertension. J Cardiovasc Pharmacol 1999; 33:652–658.

33-Valkonen VP, et al. Risk of acute coronary events and serum concentration of asymmetrical dimethylarginine. Lancet 2001; 358:2127–2128.

10-Role of L-arginine in erectile dysfunction and infertility

Male erectile dysfunction has been defined as the persistent inability to attain and maintain an erection adequate to permit satisfactory sexual intercourse. This condition is common with 25% of males aged 45–70 years reporting moderate erectile dysfunction and 10% reporting severe erectile dysfunction [1]. Vascular disease and diabetes are well recognized as the leading causes of organic erectile dysfunction and a significant proportion of these patients have either overt or covert coronary artery disease.

Male erectile dysfunction and cardiovascular disease share many of the same risk factors, namely hypertension, diabetes, hyperlipidemia and smoking [1]. Therefore, many patients with male erectile dysfunction may be treated with anti-anginal drugs including organic nitrates. Therefore, any interaction of treatments for male erectile dysfunction with organic nitrates could lead to serious adverse events.

In a small, uncontrolled trial, men with ED were given 2.8 g arginine daily for two weeks. Forty percent of men in the treatment group experienced improvement, compared to none in the placebo group [2].

In a larger double-blind trial, men with ED were given 1,670 mg arginine daily or a matching placebo for six weeks [3]. Arginine supplementation was effective at improving ED in men with abnormal nitric oxide metabolism. However, another double-blind trial of arginine for ED (500 mg three times daily for 17 days) found the amino acid no more effective than placebo [4].

Yohimbine, a selective a 2-adrenergic receptor antagonist, may have some benefit in the treatment of male erectile dysfunction [5], but its efficacy is modest at best when given alone (for review, see Tam *et al.* [6]). Nitric oxide (NO), which is synthesized from the amino acid L -arginine, is essential for normal erections [7]. Therefore, drugs acting on the L -arginine-NO pathway are attractive as potential treatments for male erectile dysfunction. For this reason L -arginine has been combined with yohimbine as a treatment for this condition.

In double-blind, placebo-controlled studies, an on-demand oral preparation of an L -arginine/yohimbine combination significantly improved erectile function in men with mild-to-moderate disease [8, 9]. Neither L -arginine alone nor yohimbine alone was significantly better than placebo.

A similar yohimbine and L -arginine combination has been found to increase substantially vaginal pulse amplitude responses to sexual visual stimulation compared with placebo in postmenopausal women with female sexual arousal disorder [10].

Interaction of phosphodiesterase type 5 inhibitors for the treatment of erectile dysfunction with organic nitrates could lead to severe hypotension. NMI 861 is combination of 7.7 mg yohimbine tart rate and 6 g -arginine glutamate. In two placebo-controlled, randomized, double-blind, two-way crossover design studies single oral dose of NMI 861 administered in 16 healthy male subjects, and then the pharmacodynamics of orally administered NMI 861 in combination with intravenous nitroglycerine (GTN) in 12 healthy male subjects. The study showed that a combination of yohimbine and L-arginine has been shown to be bioavailable, safe and well tolerated. In contrast to sildenafil and other phosphodiesterase type 5 inhibitors, there is no evidence of a clinically significant adverse pharmacodynamics interaction between intravenous nitrates and yohimbine and L-arginine in healthy subjects under the experimental conditions in which the study was conducted [11].

Arginine is required for normal spermatogenesis. Feeding an arginine-deficient diet to adult men for nine days decreased sperm counts by approximately 90 percent and increased the percentage of non-motile sperm approximately 10-fold. Oral administration of 500 mg arginine-HCl per day to infertile men for 6-8 weeks markedly increased sperm count and motility in a majority of patients, and resulted in successful pregnancies. Similar effects on oligospermia and conception rates have been reported in other preliminary trials. However, when baseline sperm counts were less than 10 million/mL, arginine supplementation produced little or no improvement [12- 18].

In female infertility, supplementation with oral arginine (16 g/ day) in poor responders to in vitro fertilization improved ovarian response, endometrial receptivity, and pregnancy rate in one study [19].

References

1- Feldman HA, Goldstein I, Hatzichristou DG, Krane RJ, McKinlay JB. Impotence and its medical and psychosocial correlates: results of the Massachusetts Male Aging Study. J Urol 1994; 151: 54–61.

2-Chen J, Wollman Y, Chernichovsky T, et al. Effect of oral administration of high-dose nitric oxide donor L-arginine in men with organic erectile dysfunction: results of a double-blind, randomized study. BJU Int 1999; 83:269-273.

82. Klotz T, Mathers MJ, Braun M, et al. Effectiveness of oral L-arginine in first-line treatment of erectile dysfunction in a controlled crossover study. Urol Int 1999; 63:220-223.

5- Ernst E, Pittler MH. Yohimbine for erectile dysfunction: a systematic review and meta-analysis of randomized clinical trials. J Urol 1998; 159: 433–6.

6- Tam SW, Worcel M, Wyllie M. Yohimbine: a clinical review. Pharmacol Ther 2001; 91: 215–43.

7- Ignarro LJ, Cirino G, Casini A, Napoli C. Nitric oxide as a signaling molecule in the vascular system: an overview. J Cardiovascular Pharmacol 1999; 34: 879–86.

8- Lebret T, Herve JM, Gorny P, Worcel M, Botto H. Efficacy and safety of a novel combination of L-arginine glutamate and yohimbine hydrochloride: a new oral therapy for erectile dysfunction. Eur Urol 2002; 41: 608–13.

9- Padma-Nathan H. Hemodynamic effects of the eral administration of a combination of arginine and yohimbine measured by color duplex ultrasonography in men with erectile dysfunction (ED). Third Meeting of the European Society for Impotence Research (ESIR) 2000; 28.

10- Meston CM, Worcel M. The effects of yohimbine plus 1-arginine glutamate on sexual arousal in postmenopausal women with sexual arousal disorder. Arch Sexual Behav 2002; 31: 323–32.

11- Kernohan AFB, McIntyre M, Hughes DM, et al. An oral yohimbine/ L -arginine combination (NMI 861) for the treatment of male erectile dysfunction: a pharmacokinetic, pharmacodynamic and interaction study with intravenous nitroglycerine in healthy male subjects. Br J Clin Pharmacol 2004; 59:1 85–93.

12- Holt LE Jr, Albanese AA. Observations on amino acid deficiencies in man. Trans Assoc Am Physicians 1944; 58:143-156.

13- Tanimura J. Studies on arginine in human semen. Part II. The effects of medication with L-arginine- HCl on male infertility. Bull Osaka Med School 1967; 13:84-89.

14- De Aloysio D, Mantuano R, Mauloni R, Nicoletti G. The clinical use of arginine aspartate in male infertility. Acta Eur Fertil 1982; 13:133-167.

15- Scibona M, Meschini P, Capparelli S, et al. L-arginine and male infertility. Minerva Urol Nefrol 1994; 46:251-253.

16- Schacter A, Goldman JA, Zukerman Z. Treatment of oligospermia with the amino acid arginine. J Urol 1973; 110:311-313.

17- Schacter A, Friedman S, Goldman JA, Eckerling B. Treatment of oligospermia with the amino acid arginine. Int J Gynaecol Obstet 1973; 11:206-209.

18- Pryor JP, Blandy JP, Evans P, et al. Controlled clinical trial of arginine for infertile men with oligozoospermia. Br J Urol 1978; 50:47-50. 91. Mroueh A. Effect of arginine on oligospermia. Fertil Steril 1970; 21:217-219.

19-Battaglia C, Salvatori M, Maxia N, et al. Adjuvant L-arginine treatment for in vitro fertilization in poor responder patients. Hum Reprod 1999; 14:1690- 1697.

11-L-arginine in Human Immunodeficiency Virus (HIV) Infection

Arginine may be of benefit in individuals with HIV/AIDS. In a small pilot study of arginine supplementation in 11 patients with HIV, given 19.6 g/day arginine or placebo for 14 days. NK cell cytotoxicity increased 18.9 lytic units, compared to an increase of 0.3 lytic units with placebo. This was not statistically significant, most likely due to the small number of patients in the study [1].

A combination of glutamine, arginine, and hydroxymethylbutyrate (HMB) may prevent loss of lean body mass in individuals with AIDS cachexia. In a double-blind trial, AIDS patients with documented weight loss of at least five percent in the previous three months received either placebo or a combination of 3 g HMB, 14 g L-glutamine, and 14 g arginine given in two divided doses daily for eight weeks. At eight weeks, subjects consuming the mixture gained 3.0 ± 0.5 kg, while those supplemented with placebo gained only 0.37 ± 0.84 kg (p = 0.009). The weight gain in the supplemented group was predominately lean muscle mass, while the placebo group lost lean mass [2].

A six-month, randomized, double-blind trial of an arginine/essential fatty acid combination was undertaken in patients with HIV. Patients received a daily oral nutritional supplement (606 kcal supplemented with vitamins, minerals, and trace elements).Half of the patients were randomized to receive 7.4 g arginine plus 1.7 g omega-3 fatty acids daily. Body weight increased similarly in both groups, and there was no change in immunological parameters [3].

References

1-Swanson B, Keithley JK, Zeller JM, Sha BE. A pilot study of the safety and efficacy of supplemental arginine to enhance immune function in persons with HIV/AIDS. Nutrition 2002; 18:688- 690.

2- Clark RH, Feleke G, Din M, et al. Nutritional treatment for acquired immunodeficiency virus-associated wasting using beta-hydroxy beta-methyl butyrate, glutamine, and arginine: a randomized, double-blind, placebo-controlled study. JPEN J Parenter Enteral Nutr 2000; 24:133- 139.

3- Pichard C, Sudre P, Karsegard V, et al. A randomized double-blind controlled study of 6 months of oral nutritional supplementation with arginine and omega-3 fatty acids in HIV infected patients. Swiss HIV Cohort Study. AIDS 1998; 12:53-63.

12-L-arginine and cancer

Animal research has shown large doses of arginine may interfere with tumor induction [1].Short-term arginine supplementation may assist in maintenance of immune function during chemotherapy.arginine supplementation (30 g/day for three days) reduced chemotherapy-induced suppression of lymphokine-activated killer cell cytotoxicity and lymphocyte mitogenic reactivity in patients with locally advanced breast cancer.18,19 In another study, arginine supplementation (30 g/day for three days prior to surgery) significantly enhanced the activity of tumor infiltrating lymphocytes in human colorectal cancers in vivo[2].Arginine, RNA, and fish oil have been combined to improve immune function in cancer patients[3,4,5]. Arginine has also promoted cancer growth in animal and human research [6].

Polyamines act as growth factors for cancers. In several types of cancer, drugs are being investigated to inhibit ornithine decarboxylase (ODC), and hence inhibit polyamine formation. The possibility of arginine stimulating polyamine formation might be a concern in chronic administration, since both arginase and ODC appear to be up-regulated in some cancers.

References

1- Takeda Y, Tominga T, Tei N, et al. Inhibitory effect of L-arginine on growth of rat mammary tumors induced by 7, 12, dimethlybenz(a)anthracine. Cancer Res 1975; 35:390-393.
2- Heys SD, Segar A, Payne S, et al. Dietary supplementation with L-arginine: modulation of tumour-infiltrating lymphocytes in patients with colorectal cancer. Br J Surg 1997; 84:238-241.
3- Kemen M, Senkal M, Homann HH, et al. Early postoperative enteral nutrition with arginineomega-3 fatty acids and ribonucleic acid supplemented diet versus placebo in cancer patients: an immunologic evaluation of Impact. Crit Care Med 1995; 23:652-659.
4- Gianotti L, Braga M, Fortis C, et al. A prospective, randomized clinical trial on perioperative feeding with an arginine-, omega-3 fatty acid-, and RNAenriched enteral diet: effect on host response and nutritional status. JPEN J Parenter Enteral Nutr 1999; 23:314-320.
5- van Bokhorst-De Van Der Schueren MA, Quak JJ, von Blomberg-van der Flier BM, et al. Effect of perioperative nutrition, with and without arginine supplementation, on nutritional status, immune function, postoperative morbidity, and survival in severely malnourished head and neck cancer patients. Am J Clin Nutr 2001; 73:323-332.
6- Park KGM. The Sir David Cuthbertson Medal Lecture 1992. The immunological and metabolic effects of L-arginine in human cancer. Proc Nutr Soc 1993; 52:387-401.

13-Burns, trauma and perioperative nutrition

Burn injuries significantly increase arginine oxidation and can result in depletion of arginine reserves. Total parenteral nutrition (TPN) increases conversion of arginine to ornithine and proportionally increases irreversible arginine oxidation, which, coupled with limited de novo synthesis from its immediate precursors, makes arginine conditionally essential in severely burned patients receiving TPN [1].

Several trials have demonstrated reduced length of hospital stay, fewer acquired infections, and improved immune function among burn 60 and trauma 61 patients supplemented with various combinations of fish or canola oil, nucleotides, and arginine.

Arginine is a potent immunomodulator. Evidence suggests a beneficial effect of arginine supplementation in catabolic conditions such as sepsis and postoperative stress. Supplemental arginine appears to up-regulate immune function and reduce the incidence of postoperative infection [2].

Two controlled trials have demonstrated increased lymphocyte mitogenesis and improved wound healing in experimental surgical wounds in volunteers given 17-25 g oral arginine daily. [3, 4]. Similar results have been obtained in healthy elderly volunteers[5].

References

1- Yu YM, Ryan CM, Castillo L, et al. Arginine and ornithine kinetics in severely burned patients: increased rate of arginine disposal. Am J Physiol Endocrinol Metab 2001; 280:E509-E517.
2- Evoy D, Lieberman MD, Fahey TJ 3rd, Daly JM. Immunonutrition: the role of arginine. *Nutrition* 1998; 14:611-617.
3- Barbul A, Rettura G, Levenson SM, et al. Wound healing and thymotropic effects of arginine: a pituitary mechanism of action. *Am J Clin Nutr* 1983; 37:786-794.
4- Barbul A, Lazarou SA, Efron DT, et al. Arginine enhances wound healing and lymphocyte immune responses in humans. *Surgery* 1990; 108:331-337.
5- Kirk SJ, Hurson M, Regan MC, et al. Arginine stimulates wound healing and immune function in elderly human beings. *Surgery* 1993; 114:155-160.

14-L-arginine in miscellaneous conditions

1-Smoking

Cigarette smoking is associated with increased monocyte endothelial
cell adhesion when endothelial cells are exposed to serum from healthy
young adults. This abnormality is acutely reversible by oral L-Arginine but
not by vitamin C.

Cigarette smoking has been associated with abnormal endothelial function
and increased leukocyte adhesion to endothelium, both key early events in
atherogenesis. Supplementation with both oral L-arginine (the physiologic
substrate for nitric oxide) and vitamin C (an aqueous phase antioxidant) may
improve endothelial function.

On study showed that in smokers compared with control subjects, monocyte/
endothelial cell adhesion was increased, endothelial expression of
intercellular adhesion molecule (ICAM)-1 was increased, and vitamin C
levels were reduced. After oral L-arginine, monocyte/ endothelial cell
adhesion was reduced in smokers, as was endothelial cell expression of
ICAM-1. After vitamin C, there was no significant change in monocyte/
endothelial cell adhesion or ICAM-1 expression from baseline in the
smokers despite an increase in vitamin C levels (to 115 +/- 7 mmol/liter) [1]

2-Preterm Labor and Delivery

Evidence from human and animal studies indicates nitric oxide inhibits
uterine contractility and may help maintain uterine quiescence during
pregnancy [2].IV arginine infusion (30 g over 30 min) in women with
premature uterine contractions transiently reduced uterine
contractility[3].Further research is needed to confirm the efficacy and safety
of arginine in prevention of preterm delivery.

3-Gastrointestinal Conditions

Preliminary evidence suggests arginine accelerates ulcer healing due to its
hyperemic, angiogenic, and growth-promoting actions, possibly involving
NO, gastrin, and polyamines [4, 5]. No clinical trials have yet explored the
efficacy of arginine supplementation as a treatment for gastritis or peptic
ulcer in humans.

A small, double-blind trial found oral arginine supplementation significantly decreased the frequency and intensity of chest pain attacks in patients with esophageal motility disorders [6].In another study, arginine infusions (500 mg/kg body weight/120 min) failed to affect lower esophageal sphincter motility.

4-Interstitial Cystitis (IC)

In an uncontrolled trial, 10 patients with IC took 1.5 g arginine daily for six months. Supplementation resulted in a significant decrease in urinary voiding discomfort, lower abdominal pain, and vaginal/urethral pain. Urinary frequency during the day and night also significantly decreased[7].In a five week uncontrolled trial, however, arginine supplementation was not effective, even at higher doses of 3-10 g daily[8].In a randomized, double-blind trial of arginine for IC, patients took 1.5 g arginine daily for three months. Twenty-nine percent of patients in the arginine group and eight percent in the placebo group experienced clinical improvement (i.e., decreased pain and urgency) by the end of the trial (p = 0.07). The results fell short of statistical significance, most likely because of the small sample size (n = 53).

References

1-Adams MR, Jessup W, Celermajer DS. Cigarette Smoking is associated with Increased Human Monocyte Adhesion to Endothelial Cells: Reversibility with Oral L-Arginine but not Vitamin C J Am Coll Cardiol. 1995 Nov 1; 26(5):1251-6.

2- Buhimschi IA, Saade GR, Chwalisz K, Garfield RE. The nitric oxide pathway in pre-eclampsia: pathophysiological implications. Human Reprod Update 1998; 4:25-42.

3- Facchinetti F, Neri I, Genazzani AR. L-arginine infusion reduces preterm uterine contractions. J Perinat Med 1996; 24:283-285.

4- Brzozowski T, Konturek SJ, Sliwowski Z, et al. Role of L-arginine, a substrate for nitric oxidesynthase, in gastroprotection and ulcer healing. J Gastroenterol 1997; 32:442-452. 5- Brzozowski T, Konturek SJ, Drozdowicz D, et al. Healing of chronic gastric ulcerations by L-Arginine. Role of nitric oxide, prostaglandins, gastrin and polyamines. Digestion 1995; 56:463-471.

6-.Bortolotti M, Brunelli F, Sarti P, Miglioli M. Clinical and manometric effects of L-arginine in patients with chest pain and oesophageal motor disorders. Ital J Gastroenterol Hepatol 1997; 29:320-324.

7- Smith SD, Wheeler MA, Foster HE Jr, Weiss RM. Improvement in interstitial cystitis symptom scores during treatment with oral L-arginine. J Urol 1997; 158:703-708.

8- Ehren I, Lundberg JO, Adolfsson J. Effects of L-arginine treatment on symptoms and bladder nitric oxide levels in patients with interstitial cystitis. Urology 1998; 52:1026-1029.

15-L-Arginine supplementation

Dosage

Doses of arginine used in clinical research have varied considerably, from as little as 500 mg/day for oligospermia to as much as 30 g/day for cancer, preeclampsia, and premature uterine contractions.

Typical daily doses fall into either the 1-3 g or 7-15 g range, depending on the condition being treated. Because of the pharmacokinetics of L-arginine, use of a sustained-release preparation may be preferable, in order to keep blood levels more constant over time.

Side Effects and Toxicity

Arginine has been shown to be safe many studies.Arginine has been used safely in humans for the past 30 years [1].
Significant adverse effects have not been observed with arginine supplementation. People with renal failure or hepatic disease may be unable to appropriately metabolize and excrete supplemental arginine and should be closely monitored when taking arginine supplements.

Contraindications

It has been postulated, on the basis of older in vitro data [2] and anecdotal reporting, that arginine supplementation might be contraindicated in persons with herpes infections (i.e., cold sores, genital herpes). The assumption is that arginine might stimulate replication of the virus and/or provoke an outbreak; however, this caution has not been validated by controlled clinical trials.

Bronchoconstriction is reportedly inhibited by the formation of NO in the airways of asthmatic patients, and a bronchoprotective effect of NO in asthma has been proposed [3]. Airway obstruction in asthma might be associated with endogenous NO deficiency caused by limited availability of NO synthase substrate (i.e., arginine).

Oral arginine (50 mg/kg body weight) in asthmatic patients triggered by a histamine challenge produced only a marginal, statistically insignificant improvement of airway hyper- responsiveness to histamine [4]. In fact, it is unclear whether NO acts as a protective or a stimulatory factor in airway hyper-responsiveness.

Since polyamines act as growth factors for cancers, and arginine may stimulate polyamine synthesis, chronic administration of arginine in cancer patients should probably be avoided until information arises regarding the safety of this practice.

Reference

1-Fideieff HL, et al. Reproducibility and safety of the arginine test in normal adults. Medicina (B Aires) 1999; 59(3):249-53.
2-Tankersley RW. Amino acid requirements of herpes simplex virus in human cells. J Bacteriol 1964; 87:609-613.
3-Ricciardolo FL, Geppetti P, Mistretta A, et al. Randomised double-blind placebo-controlled study of the effect of inhibition of nitric oxide synthesis in bradykinin-induced asthma. Lancet 1996; 348:374-377.
4- de Gouw HW, Verbruggen MB, Twiss IM, Sterk PJ. Effect of oral L-arginine on airway hyperresponsiveness to histamine in asthma. Thorax 1999; 54:1033-1035.

16-Therapeutic potentials and health benefits of L-arginine

There are a huge number of scientific researches suggesting health benefits of L-arginine supplementation and an important therapeutic potential in many conditions [1-88]:

1-L-Arginine inhibits one of the primary mechanisms of the aging process (it inhibits the process of cross-linking) and increases the release of the human growth hormone (HGH) (also known as the anti-aging hormone) from the pituitary gland. It also exerts antioxidant effects that scavenge superoxide free radicals. [1, 2, 48, 51].

2-Arginine helps to prevent atherosclerosis and reduces the severity of existing atherosclerosis and inhibits the adhesion of monocytes to the endothelium (an underlying event in the course of atherosclerosis).L-Arginine lowers total serum cholesterol levels, low-density lipoprotein (LDL) levels elevated serum triglyceride levels., and restores normal endothelial function in hypercholesterolemia[16,49,50,60].It improves blood circulation (by stimulating the production of nitric oxide, an endogenous neurotransmitter that helps to prevent vasoconstriction and which initiates vasodilatation by relaxing the smooth muscle cells of the blood vessels). Arginine helps in preventing abnormal blood clotting (by stimulating the production of plasmin and by increasing vasodilatation).Arginine also helps in preventing free radicals-induced damage to the lining of blood vessels (by enhancing the production of nitric oxide in blood vessels) [4-8].

3-L-Arginine reverses consequences of coronary heart disease [10] and improves blood circulation, improves exercise capability and facilitates vasodilatation in angina patients [3, 21].

4- L-Arginine improves heart failure as it significantly increases stroke volume and cardiac output (without effect on heartbeat rate) in congestive heart failure patients. It also increases vasodilatation (leading to increased blood circulation) in congestive heart failure patients [9, 22].

5-L-Arginine lowers blood pressure in some hypertension patients (by facilitating the body's production of nitric oxide (NO) and by inhibiting the angiotensin converting enzyme (ACE)) and reverses adverse effects of high blood pressure [11-13].

6-L-Arginine reduces pulmonary blood pressure and improves blood circulation in pulmonary hypertension Patients [14].

7-L-Arginine increases walking distance in intermittent Claudication patients and improve walking distance in peripheral vascular disease [15, 17, 23].

8-L-Arginine improves outcome after bypass surgery and helps in preventing restenosis after angioplasty and bypass [18-20]

9-L-Arginine deficiency can cause constipation and its supplementation may decrease the incidence of gallstones and improve irritable bowel syndrome (IBS). L- arginine may promote the healing of associated with ulcerative colitis patients. L-Arginine also helps in prevent post surgical damage after intestinal manipulation [24-28].

10-L-Arginine alleviates the pain and discomfort associated with interstitial cystitis and significantly improves the function of the kidneys and prevents age-related degradation of the kidneys [30-31].

11- Arginine helps to prevent bacterial and viral diseases in persons with suppressed immune systems and improves outcome in sepsis [29,32].It stimulates numerous components of the immune system such as the production of helper T-cells, the activity and production of lymphocytes by the thymus gland, the activity (cytotoxicity) of NK lymphocytes, the production of T-lymphocytes within the thymus and makes them more active and effective, the size of the thymus, and also stimulates the production of lymphocytes by the thymus and restores the production of thymic hormones to youthful levels[38-43].

12-L-Arginine inhibits the cellular replication of 24 different types of cancer in animals and boosts the ability of the immune system to fight breast cancer it also lowers tumor protein synthesis and tumor growth rate in liver cancer and inhibits the growth of some types of sarcomas.-arginine reduce the activity of ornithine decarboxylase, an enzyme that is associated with some types of cancer and may improve the outcome of cancer treatment [33-37].

13-L-Arginine may prevent diabetes. It may reduce insulin resistance and improves blood sugar disposal in diabetes and also reverses damage caused by diabetes [52-55].

14-L-Arginine increases oxygen uptake in the lungs in persons with hypoxia and increases oxygen uptake in the lungs in persons with altitude sickness and may improve asthma [56-58].

15- Arginine helps to detoxify the liver and alleviates cirrhosis. Liver malfunction can occur as a result of arginine deficiency [59].

16- L-Arginine alleviates obesity and facilitates weight loss (by stimulating the release of human growth hormone (HGH) from the pituitary gland) [61].

17-L- Arginine facilitates the healing of fractures and muscle growth (by inhibiting muscle loss) and is required for the transport of the nitrogen used in muscle metabolism. L-Arginine may prevent and alleviate osteoporosis (by stimulating the release of human growth hormone (HGH) .Muscle weakness can occur as a result of arginine deficiency. As a precursor for nitric oxide production, Arginine causes the relaxation of smooth muscle. L-Arginine improves muscle performance and improves glucose uptake into muscle cells [62-68].

18-L- Arginine is essential for the regeneration of damaged axons of neurons (its role appears to be as an agent for degrading proteins that have been damaged through axon injury).It facilitates the potentiation of long-term memory (by stimulating the production of nitric oxide (NO) - a neurotransmitter responsible for the potentiation (storage) of long-term memory and also improves memory and cognitive functions and improves pituitary responsiveness and modulates hormonal control. It may be useful for the treatment of Alzheimer's disease (due to its ability to repair damaged axons by increasing polyamines levels) [69-73].

19-L-Arginine improve impotence and male infertility by improving sperm count and sperm motility (due to its involvement in the manufacture of endogenous spermidine).It also enhances (male and female) sexual desire (libido) (female) sexual performance - due to its role in the production of nitric oxide in the clitoris (nitric oxide facilitates female orgasm in the clitoris).L-Arginine also improves (male) sexual performance by providing nitrogen to the nitric oxide (NO) molecule that is integral to the achievement of erections .Arginine produces erections that are bigger, harder and more frequent. It also increases male sexual endurance, i.e. erections that last for a longer period of time. L-arginine may improve prostate function and its deficiency can cause atrophy of the testicles of the testes [74-80].

78

20- Hair loss (especially male pattern baldness) can occur as a result of Arginine deficiency [81].

20-L-Arginine concentrates in the skin and when applied topically increases the level of vascular endothelial growth factor in the skin. It stimulates the proliferation of fibroblasts (skin cells) and is essential for and accelerates the healing of wounds (by stimulating the release of human growth hormone (HGH), stimulating the production of collagen and by stimulating the proliferation of fibroblasts).L-arginine accelerates the healing of burns and dramatically accelerates the healing of wounds in people who have undergone surgery. It decreases post operative infection and length of hospital stay [82-87]. Arginine helps to counteract inflammation and accelerates the ability of the immune system to recover from surgery. [44, 45] .L-Arginine may also improve scleroderma [88].

21- L-Arginine reduces blood clots and strokes [19].

22-Arginine improves sickle cell disease [46].

23-Alkalosis can occur as a result of arginine deficiency [47].

References

1. Radner W, et al. L-arginine reduces kidney collagen accumulation and N-epsilon-(carboxymethyl) lysine in the aging NMRI-mouse. J Gerontol 1994; 49(2):M44-M46.

2. Gianotti L, Macario M, Lanfranco F, et al. Arginine counteracts the inhibitory effect of recombinant human insulin-like growth factor I on the somatotroph responsiveness to growth hormone-releasing hormone in humans. J Clin Endocrinol Metab 2000; 85(10):3604-8.

3. Ceremuzynski L, et al. Effect of supplemental oral L-arginine on exercise capacity in patients with stable angina pectoris. Am J Cardiol 1997; 80:331-333.

4. Adams, R. R., et al. Oral L-arginine improves endothelium-dependent dilatation and reduces monocyte adhesion to endothelial cells in young men with coronary artery disease. Atherosclerosis 1997; 129(2):261-269.

5. Adams M R., et al. Cigarette smoking is associated with increased human monocyte adhesion to endothelial cells: reversibility with oral L-arginine but not vitamin C. Journal of the American College of Cardiology 1997; 29(3):491-497.

6. Huk I, et al. L-arginine treatment alters the kinetics of nitric oxide and superoxide release and reduces ischemica/reperfusion injury in skeletal muscle. Circulation 1997; 96:667-675.

7. Drexler H, et al. Correction of endothelial dysfunction in coronary microcirculation of hypercholesterolaemic patients by L-arginine. The Lancet 1991; 338:1546-50.

8. Huk I, et al. L-arginine treatment alters the kinetics of nitric oxide and superoxide release and reduces ischemica/reperfusion injury in skeletal muscle. Circulation 1997; 96:667-675.

9. Koifman B, et al. Improvement of cardiac performance by intravenous infusion of l-arginine in patients with moderate congestive heart failure. Journal of the American College of Cardiology. 26(5):1251-6, 1995.

10. Quyyumi AA. Does acute improvement of endothelial dysfunction in coronary artery disease improve myocardial ischemia? J Am Coll Cardiol 1998; 32(4):904-11.

11. Khosh, F. Natural approach to hypertension. Alternative Medicine Review 2001; 6(6), 2001.

12. Sisic D, Francishetti A, Frolich ED. Prolonged L-arginine on cardiovascular mass and myocardial hemodynamics and collagen in aged spontaneously hypertensive and normal rats. Hypertension 1999; 33(1 Pt 2):451-5.

13. Nakaki T, et al. L-arginine induced hypotension. Lancet 1990Oct 20; 336(8721):1016-7.

14. Nagaya N., et al. Short-term oral administration of L-arginine improves hemodynamics and exercise capacity in patients with precapillary pulmonary hypertension. Am J Resp Crit Care Med 2001; 163(4):887-891.

15. Roberts A. J., et al. Nutraceuticals: The Complete Encyclopedia of Supplements, Herbs, Vitamins and Healing Foods. Berkely Publishing Group. New York, USA. 2001:319.

16. Maxwell AJ, Anderson B Zapien MP, Cooke JP. Endothelial dysfunction in hypercholesterolemia is reversed by: nutritional product designed to enhance nitric oxide activity. Cardiovasc Drugs Ther 2000; 14(3):309-16.

17. Maxwell AJ, Anderson BE Cooke JP. Nutritional therapy for peripheral artery disease. Vasc Med 2000; 5(1):11-19.

18. Wallace AW, Ratcliffe MB, Galindez D, Kong JS. L-arginine infusion dilates coronary vasculature in patients undergoing coronary bypass surgery Aenesthesiology 1999; 90(6):1577-8.

19. Bode-Boger SM, Boger RH, et al. Differential inhibition of human platelet aggregation and thromboxane A 2 formation by L-arginine in vivo and in vitro. Arch Pharmacol 1998; 357:143-150.

20. Le Yorneau T, Van Belle E, Corseaux D, et al. Role of nitric oxide in re-stenosis after experimental balloon angioplasty in the hypercholesterolemic rabbit. J Am CollCardiol 1999; 33(3):876-82.

21. Suematsu Y, Ohtsuka T, et al. L-Arginine given after ischemic preconditioning can enhance cardioprotection in isolated rat hearts. Eur J Cardiothorac Surg 2001, Jun; 19(6):873-9.

22. Hambrecht R, et al. Correction of endothelial dysfunction in chronic heart failure: additional effects of exercise training and oral L-arginine supplementation. J Am Coll Cardiol 2000; 35(3):706-13.

23. Bode-Boger SM, Boger RH, et al. L-arginine induces nitric oxide-dependent vasodilation in patients with critical limb ischemia. A randomized, controlled study. Circulation 1996; 93(1):85-90.

24. Miller, A. L. The pathogenesis, clinical implications, and treatment of intestinal hyperpermeability. Alternative Medicine Review 1997; 2(5):330-345.

25. Segala, M. (editor). Disease Prevention and Treatment 3rd Edition. Life Extension Media. Florida, USA. 2000:202.

26. Sahin AS, Atalik KE, Gunel E, Dogan N. Nonadrenergic, noncholinergic responses of the human colon smooth muscle and the role of K+channels in these responses. Methods Find Exp Clin Pharmacol 2001; 23(1):13-7.

27. Khattab MM, Gad MZ, Abdallah D. Protective role of nitric oxide in indomethacin- induced gastric ulceration by a mechanism independent of gastric acid secretion. Pharmacol Res 2001; 43(5):463-7.

28. Thomas S, Ramachandran A, Patra S, et al. Nitric oxide protects the intestine from the damage induced by laparotomy and gut manipulation. J Surg Res 2001; 99(1):25-32.

29. Vallet B. Microthrombosis in sepsis. Minerva Anestesiol 2001; 67(4):298-301.

30. Smith, S. D., et al. Improvement in interstitial cystitis symptoms scores during treatment with oral L-arginine. J Urol 1997; 158(3 Part 1):703-708.

31. Reckelhoff, J. F., et al. Long-term dietary supplementation with L-arginine prevents age-related reduction in renal function. Am J Physiol. 1997; 272(6 Part 2):R1768-R1774.

32. Field, C. J., et al. Glutamine and arginine: immunonutrients for improved health. Med Sci 200; Sports Exerc. 32 :(Suppl) S377-88.

33. Reynolds, J., et al. Immunologic effects of arginine supplementation in tumor-bearing and non-tumor-bearing hosts. Annals of Surgery. 211:202-209, 19.

34. Cha-Chung, Y. Arrest of mammary tumor growth by l-arginine. Biochemical and Biophysical Research Communications 1980; 95:1306-1313.

35. Weisburger, J. Prevention by arginine glutamate of the carcinogenicity of acetamide in rats. Toxicology and Applied Pharmacology 1969 14:163-175.

36. Rettura, G., et al. Supplemental arginine increases thymic cellularity in normal and murine sarcoma virus-inoculated mice and increases the resistance to murine sarcoma virus tumour. J Par Ent Nutr 1979 3:409-416.

37. Heys SD, et al. Dietary supplementation with L-arginine: Modulation of tumor infiltrating lymphocytes in patients with colo-rectal cancer. Br J Surg 1997; 84(2):238-41.

38. Kirk, S. J., et al. Arginine stimulates wound healing and immune function in elderly human beings. Surgery 1993; 114(2):155-159.

39. Blechman S, et al. L-arginine boosts the immune system. Muscular Development 2001; 38(10):72, 2001.

40. Barbul A., et al. Arginine stimulates lymphocyte immune response in healthy human beings. Surgery 1981; 90:224-251.

41. Ochoa, J. B., et al. Effects of L-arginine on the proliferation of T lymphocyte subpopulations. J Parenteral Enteral Nutr 2001; 25:23-29.

42. Moriguchi S., et al. Functional changes in human lymphocytes and monocytes after in vitro incubation with arginine. Nutrition Research. 7:719-729, 1987.

43. Dean, W. The neuroendocrine theory of aging part IV: the immune homeostat. Vitamin Research News. October 1999.

44. Efron, D. T., et al. Modulation of inflammation and immunity by arginine supplements. Curr Opin Clin Nutr Metab Care 1998; 1:531-538.

45. Wilmore, D. W. The effect of glutamine supplementation in patients following elective surgery and accidental injury. Journal of Nutrition 2001; 131(9 Supplement):2543S-2549S, 2001.

46. Morric CR, Kuypers FA, et al. Patterns of arginine and nitric oxide in patients with sickle cell disease with vaso-occlusive crisis and acute chest syndrome. J Ped Hemat/Onc 2000; 22(6):515-20.

47. Braverman, Eric R. The Healing Nutrients Within. Keats Publishing, New Canaan, Connecticut, USA. 1997; 221.

48. Wascher, T. C., et al. Vascular effects of L-arginine: Anything beyond a substrate for NO synthase? Biochem Biophys Res Com 1997; 234:35-38.

49. Rossitch E, Jr., et al. L-arginine normalizes endothelial function in cerebral vessels from hypercholesterolemic rabbits. Journal of Clinical Investigation 1991; 87(4):1295-1299.

50. Ryzenhov, V. E., et al. Action of arginine on the lipid and lipoprotein content in blood serum of animals. Voprosy Meditsinskoi Khimi 1984 30(6):76-80.

51. Radner, W., et al. L-arginine reduces kidney collagen accumulation and N-epsilon-(carboxymethyl) lysine in the aging NMRI-mouse. J Gerontol 1994; 49(2):M44-M46, 1994.

52. Piatti, P. M., et al. Long-term oral L-arginine administration improves peripheral and hepatic insulin sensitivity in type 2 diabetic patients. Diabetes Care 2001; 24(5):875-880.

53. Wascher, T. C., et al. Effects of low-dose L-arginine on insulin mediated vasodilation and insulin sensitivity. Eur J Clin Invest 1997; 27:690-695.

54. Giugliana D, et al. Vascular effects of acute hyperglycemia are reversed by L-arginine. Circulation 1997; 95(7):1783-90.

55. Mohan IK, Cas UN. Effects of L-arginine-nitric oxide system on chemical induced diabetes mellitus. Free Radic Biol Med 1998; 25(7):757-65.

56. Arginine improves blood flow and exercise capacity. Life Enhancement. February 2002:23-26.

57. Beall, C. M., et al. Pulmonary nitric oxide in mountain dwellers. Nature 2001; 414(6862):411-412.

58. De Gouw HW, Verbruggen MB, Twiss IM, Sterk PJ. Effect of oral L-arginine on airway hyper-responsiveness to histamine in asthma. Thorax 1999; 54(11):1033-5.

59. Moss, Ralph W. Cancer Therapy: The Independent Consumer's Guide to Non-Toxic Treatment & Prevention. Equinox Press, Brooklyn, New York, USA. 1992; 285-287.

60. Khedara A, Kawai Y Kayashita J Kato N. Feeding rats the nitric oxide synthase inhibitor, L-N (omega) nitroarginine, elevates serum triglycerides and cholesterol and lowers hepatic fatty acid oxidation. J Nutr 1996; 126(10):2563-7.

61. Gianotti L, Macario M, Lanfranco F, et al. Arginine counteracts the inhibitory effect of recombinant human insulin-like growth factor I on the somatotroph responsiveness to growth hormone-releasing hormone in humans. J Clin Endocrinol Metab 2000; 85(10):3604-8.

62. Ashish, D., et al. Nitric oxide modulates fracture healing. Journal of Bone and Mineral Research 2000; 15(2):342-351.
63. Barbul, A. Arginine: biochemistry, physiology, and therapeutic implications. J Parent Ent Nutr 1986; 10:227-238.
64. Braverman, Eric R. The Healing Nutrients Within. Keats Publishing, New Canaan, Connecticut, USA. 1997:220.
65. Visser, J. J., et al. Arginine supplementation in the prevention and treatment of osteoporosis. Medical Hypotheses 1994; 43(5):339-342.
66. Arginine improves blood flow and exercise capacity. Life Enhancement. 2002; 23-26.
67. Stevens BR, Godfrey MD, Kaminski TW, Braith RW. High intensity dynamic human muscle performance enhanced by a metabolic intervention. Med Sci Sports Exerc 2000; 32(12):2102-2104.
68. Bradley SJ, Kingwell BA, McConell GK. Nitric oxide synthase inhibition reduces leg glucose uptake but not blood flow during dynamic exercise in humans. Diabetes 1999; 48(9):1815-21.
69. Tarkowski E, et al. Intrathecal release of nitric oxide in Alzheimer's disease and vascular dementia. Dement Geriatr Cogn Disord 2000; 11(6):322-6.
70. Cestaro B. Effects of arginine, S-adenosylmethionine and polyamines on nerve regeneration. Acta Neurol Scand Suppl 1994; 154:32-41.
71. Pautler EL. The possible role and treatment of deficient microcirculation regulation in age-associated memory impairment. Med Hypotheses 1994; 42(6):363-6.
72. Pandhi P, Balakrishnan S. Cognitive dysfunction induced by phenytoin and valproate in rats: effect of nitric oxide. Indian J Physiol Pharmacol 1999; 43(3):378-82.
73. di Luigi L, Guidetti L, Pigozzi F, et al. Acute amino acid supplementation enhances pituitary responsiveness in athletes. Med Sci Sports Exerc 1999; 31(12):1748-54.
74. Chen J, et al. Effect of oral administration of high-dose nitric oxide donor L-arginine in men with organic erectile dysfunction: results of a double blind, randomized, placebo-controlled study. British Journal of Urology 1999; 83:269-273.
75. Papp G., et al. [The role of arginine and arginase activity in fertility]. Andrologia. 11:37-41, 1979.
76. Women and sex drive. Life Enhancement. December 1999.
77. Block, W. Sexual enhancement available to women too. Viagra duality: better to NO? Life Enhancement. 1998:15-18.
78. Chen J, Wollman Y, Chernichovsky T, et al. Effect of high dose nitric oxide donor L-arginine in men with organic erectile dysfunction. BJU Int 1999; 83(3):269-73.
79. Keller DW, et al. L-arginine stimulation of human sperm motility in vitro. Biol Reprod 19975; 13:154-157.
80. Aikawa K, Yokota T, et al. Endogenous nitric oxide-mediated relaxation and nitrinergic innervation in the rabbit prostate: the change with aging. Prostate 2001 Jun 15; 48(1):40-6.
81. Revolutionary treatments for baldness: The hair re-growth formulas of Peter Proctor, M. D., Ph.D. Life Extension 1997; 3(3):2-8.
82. Block, W. The science of keeping your skin young. Life Enhancement. January 1998:15-18.
83. Kirk, SJ, et al. Arginine stimulates wound healing and immune function in elderly human beings. Surgery 1993; 114(2):155-160.
84. Barbul, A., et al. Arginine: Supplemental arginine, wound healing, and thymus: Arginine-pituitary interaction. Surgical Forum 1978; 29:93.

85. Yu, Y., et al. Kinetics of plasma arginine and leucine in pediatric burn patients. American Journal of Clinical Nutrition 1996; 64(1):60-66, 1996.

86. Tepaske, R., et al. Effect of preoperative oral immune-enhancing nutritional supplement on patients at high risk of infection after cardiac surgery: a randomised placebo-controlled trial. Lancet 2001; 358:696-701.

87. Braga M, Gianotti L Raedelli G, et al. Perioperative immunonutrition in patients undergoing cancer surgery: results of a randomized double-blind phase 3 trial. Arch Surg 1999; 134(4):428-33.

88. Freedman RR, Girgis R, Mayers MD. Acute effect if nitric oxide on Raynaud's phenomenon in scleroderma. Lancet 1999 28:354; 739.

Lightning Source UK Ltd.
Milton Keynes UK
UKHW010631250321
380972UK00001B/90